TRAINEE WORKBOOK FOR CONSTRUCTION

Table of Contents

Employers Must Provide and Pay for Most PPE

Page 1

Personal Protective Equipment (PPE)

The Occupational Safety and Health Administration (OSHA) requires that employers protect you from workplace hazards that can cause injury or illness. Controlling a hazard at its source is the best way to protect workers. However, when engineering, work practice and administrative controls are not feasible or do not provide sufficient protection, employers must provide personal protective equipment (PPE) to you and ensure its use.

PPE is equipment worn to minimize exposure to a variety of hazards. Examples include items such as gloves, foot and eye protection, protective hearing protection (earplugs, muffs), hard hats and respirators.

Employer Obligations	Workers should:
✓ Performing a "hazard assessment" of the workplace to identify and control physical and health hazards.	✓ Properly wear PPE
✓ Identifying and providing appropriate PPE for employees.	✓ Attend training sessions on PPE
✓ Training employees in the use and care of the PPE.	✓ Care for, clean and maintain PPE, an
✓ Maintaining PPE, including replacing worn or damaged PPE.	✓ Inform a supervisor of the need to repair or replace PPE.
✓ Periodically reviewing, updating and evaluating the effectiveness of the PPE program.	

Employers Must Pay for Personal Protective Equipment (PPE)

On May 15, 2008, a new OSHA rule about employer payment for PPE went into effect. With few exceptions, OSHA now requires employers to pay for personal protective equipment used to comply with OSHA standards. The final rule does not create new requirements regarding what PPE employers must provide.

The standard makes clear that employers cannot require workers to provide their own PPE and the worker's use of PPE they already own must be completely voluntary. Even when a worker provides his or her own PPE, the employer must ensure that the equipment is adequate to protect the worker from hazards at the workplace.

Examples of PPE that Employers Must Pay for Include:

- Metatarsal foot protection
- Rubber boots with steel toes
- Non-prescription eye protection
- Prescription eyewear inserts/lenses for full face respirators
- Goggles and face shields
- Fire fighting PPE (helmet, gloves, boots, proximity suits, full gear)
- Hard hats
- Hearing protection
- Welding PPE

Employers Must Provide and Pay for Most PPE

Page 2

Payment Exceptions under the OSHA Rule

Employers are not required to pay for some PPE in certain circumstances:

- Non-specialty safety-toe protective footwear (including steel-toe shoes or boots) and non-specialty prescription safety eyewear provided that the employer permits such items to be worn off the job site. (OSHA based this decision on the fact that this type of equipment is very personal, is often used outside the workplace, and that it is taken by workers from jobsite to jobsite and employer to employer.)
- Everyday clothing, such as long-sleeve shirts, long pants, street shoes, and normal work boots.
- Ordinary clothing, skin creams, or other items, used solely for protection from weather, such as winter coats, jackets, gloves, parkas, rubber boots, hats, raincoats, ordinary sunglasses, and sunscreen
- Items such as hair nets and gloves worn by food workers for consumer safety.
- Lifting belts because their value in protecting the back is questionable.
- When the employee has lost or intentionally damaged the PPE and it must be replaced.

OSHA Standards that Apply

OSHA General Industry PPE Standards

- 1910.132: General requirements and payment
- 1910.133: Eye and face protection
- 1910.134: Respiratory protection
- 1910.135: Head protection
- 1910.136: Foot protection
- 1910.137: Electrical protective devices
- 1910.138: Hand protection

OSHA Construction PPE Standards

- 1926.28: Personal protective equipment
- 1926.95: Criteria for personal protective equipment
- 1926.96: Occupational foot protection
- 1926.100: Head protection
- 1926.101: Hearing protection
- 1926.102: Eye and face protection
- 1926.103: Respiratory protection

There are also PPE requirements in shipyards and marine terminals and many standards on specific hazards, such as 1910.1030: Bloodborne pathogens and 1910.146: Permit-required confined spaces.

OSHA standards are online at www.osha.gov.

Sources:

- *Employers Must Provide and Pay for PPE, New Jersey Work Environment Council (WEC) Fact Sheet*
- *OSHA Standards, 1910.132(h) and 1926.95(d)*
- *Employer Payment for Personal Protective Equipment Final Rule, Federal Register: November 15, 2007 (Volume 72, Number 220)*

Job Safety and Health IT'S THE LAW!

All workers have the right to:

- A safe workplace.
- Raise a safety or health concern with your employer or OSHA, or report a work-related injury or illness, without being retaliated against.
- Receive information and training on job hazards, including all hazardous substances in your workplace.
- Request a confidential OSHA inspection of your workplace if you believe there are unsafe or unhealthy conditions. You have the right to have a representative contact OSHA on your behalf.
- Participate (or have your representative participate) in an OSHA inspection and speak in private to the inspector.
- File a complaint with OSHA within 30 days (by phone, online or by mail) if you have been retaliated against for using your rights.
- See any OSHA citations issued to your employer.
- Request copies of your medical records, tests that measure hazards in the workplace, and the workplace injury and illness log.

This poster is available free from OSHA.

Employers must:

- Provide employees a workplace free from recognized hazards. It is illegal to retaliate against an employee for using any of their rights under the law, including raising a health and safety concern with you or with OSHA, or reporting a work-related injury or illness.
- Comply with all applicable OSHA standards.
- Notify OSHA within 8 hours of a workplace fatality or within 24 hours of any work-related inpatient hospitalization, amputation, or loss of an eye.
- Provide required training to all workers in a language and vocabulary they can understand.
- Prominently display this poster in the workplace.
- Post OSHA citations at or near the place of the alleged violations.

On-Site Consultation services are available to small and medium-sized employers, without citation or penalty, through OSHA-supported consultation programs in every state.

Contact OSHA. We can help.

1-800-321-OSHA (6742) • TTY 1-877-889-5627 • www.osha.gov

Topic 4: Workers Rights Practice Worksheet
Crossword Puzzle

OSHA Provides Workers the Right to:

Across

4. Hazard _____ and medical records
6. Information about _______ and illnesses in your workplace
7. A safe and ______ workplace
8. Complain or request hazard _____ from employer
9. Participate in an OSHA ______

Down

1. Know about _______ conditions
2. Be free from ________ for exercising safety and health rights
3. _______ as provided in the OSHA standards
5. File a complaint with _____

Filing an OSHA Complaint – Tips for Completing the OSHA-7 Form

INSTRUCTIONS Provided on the Form:
Open the form and complete the front page as accurately and completely as possible. Describe each hazard you think exists in as much detail as you can. If the hazards described in your complaint are not all in the same area, please identify where each hazard can be found at the worksite. If there is any particular evidence that supports your suspicion that a hazard exists (for instance, a recent accident or physical symptoms of employees at your site) include the information in your description. If you need more space than is provided on the form, continue on any other sheet of paper. After you have completed the form, return it to your local OSHA office.

Here are tips for completing the form:

U. S. Department of Labor
Occupational Safety and Health Administration (1)

Notice of Alleged Safety or Health Hazards

		Complaint Number	
Establishment Name			
Site Address (2)	Site Phone	Site FAX	
Mailing Address	Mail Phone	Mail FAX	
Management Official		Telephone	
Type of Business			

HAZARD DESCRIPTION/LOCATION. Describe briefly the hazard(s) which you believe exist. Include the approximate number of employees exposed to or threatened by each hazard. Specify the particular building or worksite where the alleged violation exists.

(3)

Has this condition been brought to the attention of:	☐ Employer ☐ Other Government Agency(specify) (4)
Please Indicate Your Desire:	☐ Do NOT reveal my name to my Employer ☐ My name may be revealed to the Employer (5)
The Undersigned believes that a violation of an Occupational Safety or Health standard exists which is a job safety or health hazard at the establishment named on this form.	(Mark "X" in ONE box) ☐ Employee ☐ Federal Safety and Health Committee ☐ Representative of Employees ☐ Other (specify)
Complainant Name	Telephone
Address(Street,City,State,Zip) (6)	
Signature	Date

If you are an authorized representative of employees affected by this complaint, please state the name of the organization that you represent and your title:
Organization Name: Your Title:

1. Be specific and include appropriate details: The information on the complaint form may be the only description of the hazard that the inspector will see before the inspection. The inspector will base his or her research and planning on this information.
2. Establishment Name, Address, & Type of Business: Be thorough and specific. The inspector's research on the company and the industry's hazards will be based on this information.
3. Hazard Description/Location: The hazard description is the most important part of the form. Your answers
clearly. If your complaint is about chemicals, identify them whenever possible and attach copies ofl abels or SDSs if you can. Identify the location so the inspector will know where to look.
4. Has this condition been brought to the attention of the employer or another government agency? You should indicate on the form if you have tried to get the employer to fix the hazard before filing the complaint. Also, if another agency, such as a local fire or building department, has been n otified of these hazards, OSHA may want to consult with them.
5. Do NOT reveal my name: OSHA will keep your name offthe complaint, if you wish. Remember that discrimination for health and safety activity is illegal. If you are a union representative, you may wish to have your name on the complaint.
6. Signature and address: It is important to sign the complaint if you want OSHA to conduct an onsite inspection. Also, your address will allow OSHA to send copies ofi nspection related materials to you.

Construction Complaint Scenario

Use the following scenario to determine what information should be put on an OSHA complaint form. Is any additional information needed?

You are a construction worker for ABC, Inc, 1000 Sweet Road, Anytown, USA, 40001. ABC does non-residential plumbing, heating and air-conditioning work. You have worked for ABC for three years. You, along with seven co-workers, have been installing sheetmetal ductwork in the lower level of the Anytown Shopping Mall, which is undergoing renovation, for the past few weeks. The site is located in the Northwest quadrant, in the basement of the anchor store, located at 555 Times Drive, in Anytown. One of your co-workers has been operating a 65-horsepower concrete cutting saw in the same area. The saw is being run in the propane mode. You and several co-workers get headaches from the fumes whenever the saw is used and have told your supervisor about the problem. The supervisor said that nothing could be done, because the General Contractor, CAB Management, has control over the site and this job will be complete in another month. You did some research and found out that exposure to propane in a confined, unventilated area can cause headaches, dizziness, difficulty breathing and unconsciousness. There is no ventilation or monitoring of the air in the area.

After talking to your union representative, you decide to file a complaint with OSHA.

NOTES:

U. S. Department of Labor
Occupational Safety and Health Administration

Notice of Alleged Safety or Health Hazards

For the General Public:

This form is provided for the assistance of any complainant and is not intended to constitute the exclusive means by which a complaint may be registered with the U.S. Department of Labor.

Sec 8(f)(1) of the Williams-Steiger Occupational Safety and Health Act, 29 U.S.C. 651, provides as follows: Any employees or representative of employees who believe that a violation of a safety or health standard exists that threatens physical harm, or that an imminent danger exists, may request an inspection by giving notice to the Secretary or his authorized representative of such violation or danger. Any such notice shall be reduced to writing, shall set forth with reasonable particularity the grounds for the notice, and shall be signed by the employee or representative of employees, and a copy shall be provided the employer or his agent no later than at the time of inspection, except that, upon request of the person giving such notice, his name and the names of individual employees referred to therein shall not appear in such copy or on any record published, released, or made available pursuant to subsection (g) of this section. If upon receipt of such notification the Secretary determines there are reasonable grounds to believe that such violation or danger exists, he shall make a special inspection in accordance with the provisions of this section as soon as practicable to determine if such violation or danger exists. If the Secretary determines there are no reasonable grounds to believe that a violation or danger exists, he shall notify the employees or representative of the employees in writing of such determination.

NOTE: Section 11(c) of the Act provides explicit protection for employees exercising their rights, including making safety and health complaints.

For Federal Employees:

This report format is provided to assist Federal employees or authorized representatives in registering a report of unsafe or unhealthful working conditions with the U.S.Department of Labor.

The Secretary of Labor may conduct unannounced inspection of agency workplaces when deemed necessary if an agency does not have occupational safety and health committees established in accordance with Subpart F, 29 CFR 1960; or in response to the reports of unsafe or unhealthful working conditions upon request of such agency committees under Sec. 1-3, Executive Order 12196; or in the case of a report of imminent danger when such a committee has not responded to the report as required in Sec. 1-201(h).

INSTRUCTIONS:

Open the form and complete the front page as accurately and completely as possible. Describe each hazard you think exists in as much detail as you can. If the hazards described in your complaint are not all in the same area, please identify where each hazard can be found at the worksite. If there is any particular evidence that supports your suspicion that a hazard exists (for instance, a recent accident or physical symptoms of employees at your site) include the information in your description. If you need more space than is provided on the form, continue on any other sheet of paper.

After you have completed the form, return it to your local OSHA office.

NOTE: It is unlawful to make any false statement, representation or certification in any document filed pursuant to the Occupational Safety and Health Act of 1970. Violations can be punished by a fine of not more than $10,000. or by imprisonment of not more than six months, or by both. (Section 17(g))

Public reporting burden for this voluntary collection of information is estimated to vary from 15 to 25 minutes per response with an average of 17 minutes per response, including the time for reviewing instructions, searching existing data sources, gathering and maintaining the data needed, and completing and reviewing the collection of information. An Agency may not conduct or sponsor, and persons are not required to respond to the collection of information unless it displays a valid OMB Control Number. Send comment regarding this burden estimate or any other aspect of this collection of information, including suggestions for reducing this burden to the Directorate of Enforcement Programs, Department of Labor, Room N-3119, 200 Constitution Ave., NW, Washington, DC; 20210.

OMB Approval# 1218-0064; Expires: 03-31-2011

Do not send the completed form to this Office.

OSHA-7(Rev. 9/93)

U. S. Department of Labor
Occupational Safety and Health Administration

Notice of Alleged Safety or Health Hazards

Complaint Number	

Establishment Name			
Site Address			
Site Phone		Site FAX	
Mailing Address			
Mail Phone		Mail FAX	
Management Official		Telephone	
Type of Business			

HAZARD DESCRIPTION/LOCATION. Describe briefly the hazard(s) which you believe exist. Include the approximate number of employees exposed to or threatened by each hazard. Specify the particular building or worksite where the alleged violation exists.

Has this condition been brought to the attention of:	☐ Employer ☐ Other Government Agency(specify)
Please Indicate Your Desire:	☐ Do NOT reveal my name to my Employer ☐ My name may be revealed to the Employer
The Undersigned believes that a violation of an Occupational Safety or Health standard exists which is a job safety or health hazard at the establishment named on this form.	(Mark "X" in ONE box) ☐ Employee ☐ Federal Safety and Health Committee ☐ Representative of Employees ☐ Other (specify)

Complainant Name		Telephone	
Address(Street,City,State,Zip)			
Signature		Date	

If you are an authorized representative of employees affected by this complaint, please state the name of the organization that you represent and your title:

Organization Name: Your Title:

2 OSHA-7(Rev. 3/96)

OSHA® FactSheet

OSHA's Whistleblower Protection Program

OSHA's Whistleblower Protection Program enforces the provisions of more than 20 federal laws protecting employees from retaliation for, among other things, raising or reporting concerns about hazards or violations of various workplace safety and health, aviation safety, commercial motor carrier, consumer product, environmental, financial reform, food safety, health insurance reform, motor vehicle safety, nuclear, pipeline, public transportation agency, railroad, maritime, and securities laws. Employees who believe that they have experienced retaliation in violation of one of these laws may file a complaint with OSHA.

Whistleblower Laws Enforced by OSHA

Following is a list of statutes over which OSHA has jurisdiction. Each statute has a different time frame in which a complaint can be filed.

- *Asbestos Hazard Emergency Response Act (90 days)*
- *Clean Air Act (30 days)*
- *Comprehensive Environmental Response, Compensation and Liability Act (30 days)*
- *Consumer Financial Protection Act of 2010 (180 days)*
- *Consumer Product Safety Improvement Act (180 days)*
- *Energy Reorganization Act (180 days)*
- *Federal Railroad Safety Act (180 days)*
- *Federal Water Pollution Control Act (30 days)*
- *International Safe Container Act (60 days)*
- *Moving Ahead for Progress in the 21st Century Act (motor vehicle safety) (180 days)*
- *National Transit Systems Security Act (180 days)*
- *Occupational Safety and Health Act (OSH Act) (30 days)*
- *Pipeline Safety Improvement Act (180 days)*
- *Safe Drinking Water Act (30 days)*
- *Sarbanes-Oxley Act (180 days)*
- *Seaman's Protection Act (180 days)*
- *Section 402 of the FDA Food Safety Modernization Act (180 days)*
- *Section 1558 of the Affordable Care Act (180 days)*
- *Solid Waste Disposal Act (30 days)*
- *Surface Transportation Assistance Act (180 days)*
- *Toxic Substances Control Act (30 days)*
- *Wendell H. Ford Aviation Investment and Reform Act for the 21st Century (90 days)*

What Is Retaliation?

Retaliation is an adverse action against an employee because of activity protected by one of these whistleblower laws. Retaliation can include several types of actions, such as:

- Firing or laying off
- Blacklisting
- Demoting
- Denying overtime or promotion
- Disciplining
- Denying benefits
- Failing to hire or rehire
- Intimidation
- Reassignment affecting promotion prospects
- Reducing pay or hours
- Making threats

Filing a Complaint

Employees who believe that their employers retaliated against them because they engaged in protected activity should contact OSHA as soon as possible because they must file any complaint within the legal time limits.

An employee can file a complaint with OSHA by visiting or calling his or her local OSHA office, sending a written complaint to the closest OSHA office, or filing a complaint online. No particular form is required and complaints may be submitted in any language.

Written complaints may be filed by fax, electronic communication, hand delivery during business hours, U.S. mail (confirmation services recommended), or other third-party commercial carrier.

The date of the postmark, fax, electronic communication, telephone call, hand delivery, delivery to a third-party commercial carrier, or in-person filing at an OSHA office is considered the date filed.

To file a complaint electronically, please visit: www.osha.gov/whistleblower/WBComplaint.html.

To contact an OSHA area office, employees should call 1-800-321-OSHA (6742) to be connected to the closest area office or visit www.osha.gov/html/RAmap.html to find local OSHA office address and contact information.

When OSHA receives a complaint, OSHA will first review it to determine whether certain basic requirements are met, such as whether the complaint was filed on time. If so, the complaint will be investigated in order to determine whether the employer retaliated against the employee for engaging in activity protected under one of OSHA's whistleblower laws. OSHA may also attempt to assist the employer and employee in reaching a settlement of the case.

Private-sector employees throughout the United States and its territories and employees of the United States Postal Service (USPS) who suffer retaliation because of occupational safety or health activity are covered by section 11(c) of the OSH Act. In addition, private-sector employees are also covered by laws in States which operate their own comprehensive occupational safety and health programs approved by Federal OSHA ("State Plans"). For information on the whistleblower provisions of the 22 State Plan States which cover private-sector employees, visit www.osha.gov/dcsp/osp.

With the exception of employees of the USPS, public-sector employees (those employed as municipal, county, state, territorial, or federal workers) are not covered by the OSH Act. State and local government employees are covered by the whistleblower provisions of all the States with State Plans, including six States which cover only State and local government employees.

A federal employee who is not a USPS employee who wishes to file a complaint alleging retaliation due to disclosure of a substantial and specific danger to public health or safety or involving a violation of an occupational safety or health standard or regulation should contact the Office of Special Counsel (www.osc.gov). Such federal employees are also covered by their own agency's procedures for remedying such retaliation.

Public-sector employees who are unsure whether they are covered under a whistleblower law should call 1-800-321-OSHA (6742) for assistance, or visit www.whistleblowers.gov.

Results of the Investigation

If OSHA determines that retaliation in violation of the OSH Act, *Asbestos Hazard Emergency Response Act*, or the *International Safe Container Act* has occurred, the Secretary of Labor may sue in federal district court to obtain relief. If OSHA determines that no retaliation has occurred, it will dismiss the complaint.

Under the remaining whistleblower laws, if the evidence supports an employee's complaint of retaliation, OSHA will issue an order requiring the employer, as appropriate, to put the employee back to work, pay lost wages, and provide other possible relief. If the evidence does not support the employee's complaint, OSHA will dismiss the complaint. After OSHA issues a decision, the employer and/or the employee may request a full hearing before an administrative law judge of the Department of Labor. The administrative law judge's decision may be appealed to the Department's Administrative Review Board.

Under some of the laws, an employee may file the retaliation complaint in federal district court if the Department has not issued a final decision within a specified number of days (180, 210 or 365 depending on the law).

To Get Further Information

To obtain more information on whistleblower laws, go to www.whistleblowers.gov.

This is one in a series of informational fact sheets highlighting OSHA programs, policies or standards. It does not impose any new compliance requirements. For a comprehensive list of compliance requirements of OSHA standards or regulations, refer to Title 29 of the Code of Federal Regulations. This information will be made available to sensory-impaired individuals upon request. The voice phone is (202) 693-1999; teletypewriter (TTY) number: (877) 889-5627.

Occupational Safety and Health Administration

We Are OSHA

We Can Help

Workers' rights under the OSH Act

Workers are entitled to working conditions that do not pose a risk of serious harm. To help assure a safe and healthful workplace, OSHA also provides workers with the right to:

- **Ask OSHA to inspect their workplace;**
- **Use their rights under the law without retaliation;**
- **Receive information and training about hazards, methods to prevent harm, and the OSHA standards that apply to their workplace. The training must be in a language you can understand;**
- **Get copies of test results done to find hazards in the workplace;**
- **Review records of work-related injuries and illnesses; and**
- **Get copies of their medical records.**

Occupational Safety and Health Administration
U.S. Department of Labor

1

Who OSHA covers

Private sector workers

Most employees in the nation come under OSHA's jurisdiction. OSHA covers private sector employers and employees in all 50 states, the District of Columbia, and other U.S. jurisdictions either directly through Federal OSHA or through an OSHA-approved state program. State-run health and safety programs must be at least as effective as the Federal OSHA program. To find the contact information for the OSHA Federal or State Program office nearest you, call 1-800-321-OSHA (6742) or go to www.osha.gov.

State and local government workers

Employees who work for state and local governments are not covered by Federal OSHA, but have OSH Act protections if they work in those states that have an OSHA-approved state program. The following 22 states or territories have OSHA-approved programs:

Alaska	Arizona	California
Hawaii	Indiana	Iowa
Kentucky	Maryland	Michigan
Minnesota	Nevada	New Mexico
North Carolina	Oregon	South Carolina
Tennessee	Utah	Vermont
Virginia	Washington	Wyoming
Puerto Rico		

Five additional states and one U.S. territory have OSHA-approved plans that cover public sector workers only:

Connecticut	Illinois	Maine
New Jersey	New York	Virgin Islands

Private sector workers in these five states and the Virgin Islands are covered by Federal OSHA.

2

Federal government workers
Federal agencies must have a safety and health program that meets the same standards as private employers. Although OSHA does not fine federal agencies, it does monitor federal agencies and responds to workers' complaints. The United States Postal Service (USPS) is covered by OSHA.

Not covered under the OSH Act:
- Self-employed;
- Immediate family members of farm employers who do not employ outside employees;
- Workplace hazards regulated by another federal agency (for example, the Mine Safety and Health Administration, the Department of Energy, or Coast Guard).

OSHA standards: Protection on the job

OSHA standards are rules that describe the methods that employers must use to protect their employees from hazards. There are OSHA standards for Construction work, Agriculture, Maritime operations, and General Industry, which are the standards that apply to most worksites. These standards limit the amount of hazardous chemicals workers can be exposed to, require the use of certain safe practices and equipment, and require employers to monitor hazards and keep records of workplace injuries and illnesses.

3

Examples of OSHA standards include requirements to provide fall protection, prevent trenching cave-ins, prevent some infectious diseases, assure that workers safely enter confined spaces, prevent exposure to harmful substances like asbestos, put guards on machines, provide respirators or other safety equipment, and provide training for certain dangerous jobs.

Employers must also comply with the General Duty Clause of the OSH Act, which *requires employers to keep their workplace free of serious recognized hazards*. This clause is generally cited when no OSHA standard applies to the hazard.

Workers can ask OSHA to inspect their workplace

Workers, or their representatives, may file a complaint and ask OSHA to inspect their workplace if they believe there is a serious hazard or that their employer is not following OSHA standards. A worker can tell OSHA not to let their employer know who filed the complaint. **It is a violation of the OSH Act for an employer to fire, demote, transfer or retaliate in any way against a worker for filing a complaint or using other OSHA rights.**

Written complaints that are signed by a worker or their representative and submitted to the closest OSHA office are more likely to result in an on-site OSHA inspection. You can call 1-800-321-OSHA (6742) to request a complaint form from your local OSHA office or visit www.osha.gov/pls/osha7/eComplaintForm.html to submit

4

the form online. Completed forms can also be faxed or mailed to the local OSHA office. Most complaints sent in online may be resolved informally over the phone with your employer.

When the OSHA inspector arrives, workers and their representatives have the right to:

- Go along on the inspection;
- Talk privately with the OSHA inspector; and
- Take part in meetings with the inspector and the employer before and after the inspection is conducted.

Where there is no union or employee representative, the OSHA inspector must talk confidentially with a reasonable number of workers during the course of the investigation.

When an inspector finds violations of OSHA standards or serious hazards, OSHA may issue citations and fines. A citation includes the methods an employer may use to fix a problem and the date by when the corrective actions must be completed. Workers only have the right to challenge the deadline for when a problem must be resolved. Employers, on the other hand, have the right to contest whether there is a violation or any other part of the citation. Workers or their representatives must notify OSHA that they want to be involved in the appeals process if the employer challenges a citation.

If you send in a complaint requesting an OSHA inspection, you have the right to find out the results of the OSHA inspection and request a review if OSHA does not issue citations.

5

Employer responsibilities

Employers have the responsibility to provide a safe workplace. **Employers MUST provide their employees with a workplace that does not have serious hazards and must follow all OSHA safety and health standards.** Employers must find and correct safety and health problems. OSHA further requires employers to try to eliminate or reduce hazards first by making changes in working conditions rather than just relying on masks, gloves, earplugs or other types of personal protective equipment. Switching to safer chemicals, implementing processes to trap harmful fumes, or using ventilation systems to clean the air are examples of effective ways to get rid of or minimize risks.

Employers **MUST** also:

- Prominently display the official OSHA *Job Safety and Health – It's the Law* poster that describes rights and responsibilities under the OSH Act. **This poster is free and can be downloaded from www.osha.gov.**
- Inform workers about chemical hazards through training, labels, alarms, color-coded systems, chemical information sheets and other methods.
- Provide safety training to workers in a language and vocabulary they can understand.
- Keep accurate records of work-related injuries and illnesses.
- Perform tests in the workplace, such as air sampling, required by some OSHA standards.
- Provide required personal protective equipment at no cost to workers.*
- Provide hearing exams or other medical tests required by OSHA standards.

6

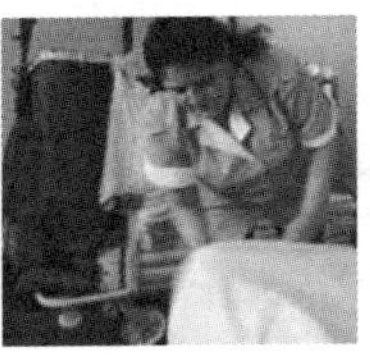

- Post OSHA citations and injury and illness data where workers can see them.
- Notify OSHA within 8 hours of a workplace fatality or within 24 hours of any work-related inpatient hospitalization, amputation or loss of an eye (1-800-321-OSHA [6742]).
- Not retaliate against workers for using their rights under the law, including their right to report a work-related injury or illness.

* Employers must pay for most types of required personal protective equipment.

The law protects workers from retaliation when using their OSHA rights

The OSH Act protects workers who complain to their employer, OSHA or other government agencies about unsafe or unhealthful working conditions in the workplace or environmental problems. You cannot be transferred, denied a raise, have your hours reduced, be fired, or punished in any other way because you used any right given to you under the OSH Act. Help is available from OSHA for whistleblowers.

If you have been punished or retaliated against for using your rights, you must file a complaint with OSHA **within 30 days** from the date the retaliatory decision was both made and communicated to you. No form is needed, but you must call OSHA within 30 days of the alleged retaliation at 1-800-321-OSHA (6742) and ask to speak to the OSHA area office nearest you to report the retaliation.

7

You have the right to a safe workplace

The *Occupational Safety and Health Act of 1970* (OSH Act) was passed to prevent workers from being killed or seriously harmed at work. The law requires that employers provide their employees with working conditions that are free of known dangers. The Act created the Occupational Safety and Health Administration (OSHA), which sets and enforces protective workplace safety and health standards. OSHA also provides information, training and assistance to workers and employers. Workers may file a complaint to have OSHA inspect their workplace if they believe that their employer is not following OSHA standards or there are serious hazards.

Contact us if you have questions or want to file a complaint. We will keep your information confidential. We are here to help you. Call our toll-free number at 1-800-321-OSHA (6742) or go to www.osha.gov.

Occupational Safety and Health Administration

U.S. Department of Labor

1-800-321-OSHA (6742) TTY 1-877-889-5627
www.osha.gov

OSHA 3334-09R 2015

8

Safety & Health Resources

Government Resources

OSHA: http://www.osha.gov/
Contact the OSHA Office nearest you or contact the toll free number:
1-800-321-OSHA (6742)

NIOSH: http://www.cdc.gov/niosh/
Phone NIOSH at
1-800-CDC-INFO (1-800-232-4636)
or Email at: cdcinfo@cdc.gov

NIOSH is a part of the Centers for Disease Control and Prevention (http://www.cdc.gov/).
CDC has extensive information on health and safety topics.

Universities

CORNELL UNIVERSITY
School of Industrial and Labor Relations:
http://www.ilr.cornell.edu/healthSafety/

LABOR OCCUPATIONAL HEALTH PROGRAM, University of California at Berkeley: http://www.lohp.org/

NATIONAL LABOR COLLEGE, George Meany Center: http://www.nlc.edu/

UCLA, Labor Occupational Safety and Health (UCLA-LOSH):
http://www.losh.ucla.edu/

COSH Groups

COSH groups are private, non-profit coalitions of labor unions, health and technical professionals, and others interested in promoting and advocating for worker health and safety. If you don't see a COSH group in your area, check the NATIONAL COSH website for local COSH groups.

NATIONAL COUNCIL FOR OCCUPATIONAL SAFETY & HEALTH National COSH is a federation of local and statewide "COSH" groups:
http://www.coshnetwork.org/

CACOSH – Chicago Area Committee on Occupational Safety and Health:
http://www.cacosh.org/

MASSCOSH –Massachusetts Coalition on Occupational Safety and Health:
http://www.masscosh.org/

NYCOSH – New York Committee for Occupational Safety and Health:
http://www.nycosh.org/

PHILAPOSH – Philadelphia Area Project for Occupational Safety and Health:
http://www.philaposh.org/
Prevention (http://www.cdc.gov/).

Unions

The following is a sample list of unions with links to useful health and safety information.
AFL-CIO: http://www.aflcio.org/issues/safety/

AFSCME: http://www.afscme.org/issues/73.cfm

eLCOSH – The Electronic Library of Construction Safety and Health is a collection of information on construction safety and health developed by CPWR – Center for Construction Research and Training, with funding by NIOSH: http://www.elcosh.org/

SEIU (Service Employees International Union) Health and Safety Department:
http://www.seiu.org/a/members/safety-and-health.php

UAW Health and Safety Department: http://www.uaw.org/hs/

Navigating the OSHA Website

http://www.osha.gov

The elements of this valuable source of occupational safety and health information are featured:

Hazard Violation
Search Workshop

29 CFR 1926 Construction Standards

Description of Hazard	Standard
1. Exhaust system for a 20-inch disc grinder cut-off wheel exhausts less than 610 ft^3/minute.	
2. Crew disposing lead-contaminated materials above PEL take lunch directly in the work area.	
3. No drinking water is present at a worksite.	
4. Spray booths are constructed from wooden frame and walls.	
5. A gas-powered generator is used near scrap lumber in an enclosed area.	
6. Fire extinguishers are obstructed.	
7. Gasoline is stored in a stairway.	
8. Portable dockboards are not secured.	
9. A natural or synthetic rope in use in rigging equipment for material handling is rotting.	
10. Oily rags are thrown away in open trashcans.	
11. A pulley in use has a broken rim.	
12. Compressed gas cylinders in transit have no valve protection cap.	
13. Oxygen pressure regulator gauges are not working.	
14. A double-throw knife switch mounted with a verticle throw lacks a locking device.	
15. Damaged scaffolding that can only hold twice the intended load is being used.	
16. Workers in an aerial lift are not wearing a body belt.	
17. The top rail of a guardrail system is 36" above the walking/working level.	
18. A hole cover in the ground is unmarked.	
19. Material hoists have no posted operating rules.	
20. Unauthorized persons are riding on a forklift.	

Hazard Violation
Search Workshop

29 CFR 1926 Construction Standards

Description of Hazard	Standard
21. A six-foot deep trench in sandy soil with workers inside has no barricade or shield system.	
22. Exposed steel rebar is protruding from the ground.	
23. A four-foot limited access zone surrounds a 12-foot masonry wall.	
24. Seven steel beams are hoisted in a single load.	
25. There is no check-in/check-out procedure for underground workers.	
26. During demolition, a debris chute opening for wheelbarrows has no toeboard.	
27. Explosives are left onsite unattended and not in a locked magazine.	
28. Live-line bare-hand work is performed on a line with an unknown voltage rating.	
29. One single-cleat ladder is provided to a second-story work area for 30 workers.	
30. A portable ladder in use has a broken rung.	
31. There is no training for workers removing asbestos.	
32. There is no compliance program for worker cadmium exposure.	
33. A crane is assembled below a live power line.	
34. A liquefied petroleum filling station for forklifts is 15 feet away from a doorway.	
35. A one-inch alloy steel chain worn one-quarter inch down is used in rigging equipment.	
36. A hammer has a cracked wooden handle.	
37. An arc welding cable is worn to bare wire.	
38. An electrical junction box is uncovered.	
39. Planking is run between window jack scaffolds.	
40. A fall arrest system in use has nine feet of slack.	

Personal Fall Arrest System Checklist

Personal Fall Arrest Systems are one way to protect workers on construction sites where there are vertical drops of six or more feet. Systems must be set up so that a worker cannot fall more than six feet, nor come into contact with any lower level.

You should be able to answer **Yes** to each of the following.

1. Is your personal fall arrest system made up of an anchorage, connecting device, and a full-body harness?

2. Are the components from the same manufacturer to ensure that the system works as it should? If not, has any substitution or change to a personal fall arrest system been fully evaluated or tested by a competent person to determine that it meets the standard?

3. Is your personal fall arrest system inspected for damage each time before you wear it? [If there are defects, or if someone has taken a fall using the equipment, it must be removed from service.]

4. Is the attachment location of the body harness in the center of your back, near the shoulder level or above your head?

5. Do vertical lifelines or lanyards have a minimum breaking strength of 5,000 lbs? Are they protected against being cut or abraded?

6. Will each worker be attached to a separate vertical lifeline?

7. Is the webbing (the materials used for ropes and straps of lifelines, lanyard, and harnesses) made of synthetic fibers?

8. Is the anchorage for your personal fall arrest equipment independent of any anchorage used to support or suspend platforms? Is it able to support at least 5,000 lbs. per worker attached to the anchorage?

9. Are the connectors made from steel or equivalent materials, with a corrosion-resistant finish and smooth edges?

10. Do the D-rings and snaphooks have a minimum tensile strength of 5,000 lbs.?

11. Are snaphooks of a locking-type and designed to prevent the snaphook from opening and slipping off the connector?

12. Are the snaphooks not *directly connected* to the webbing, rope, or wire; to each other; to a D-ring to which another snaphook or other connector is attached; to a horizontal lifeline; or to any other object that could cause the snaphook to open?

ACCIDENT SUMMARY No. 17

Accident Type:	Electrocution
Weather Conditions:	Sunny, Clear
Type of Operation:	Steel Erection
Size of Work Crew:	3
Collective Bargaining	No
Competent Safety Monitor on Site:	Yes - Victim
Safety and Health Program in Effect:	No
Was the Worksite Inspected Regularly:	Yes
Training and Education Provided:	No
Employee Job Title:	Steel Erector Foreman
Age & Sex:	43-Male
Experience at this Type of Work:	4 months
Time on Project:	4 Hours

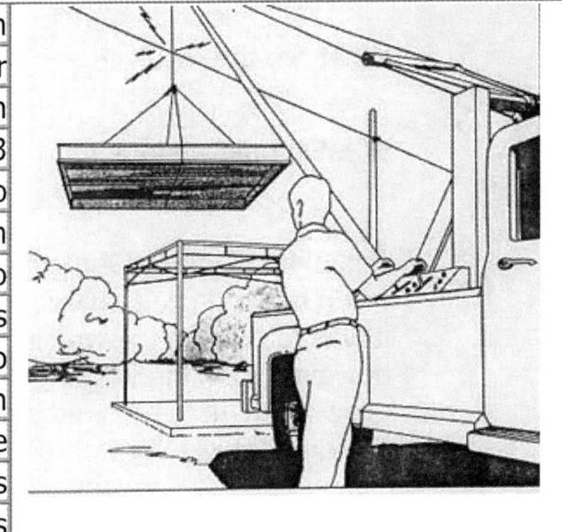

BRIEF DESCRIPTION OF ACCIDENT

Employees were moving a steel canopy structure using a "boom crane" truck. The boom cable made contact with a 7200 volt electrical power distribution line electrocuting the operator of the crane; he was the foreman at the site.

INSPECTION RESULTS

As a result of its investigation. OSHA issued citations for four serious violations of its construction standards dealing with training, protective equipment, and working too close to power lines. OSHA's construction safety standards include several requirements which, if they had been followed here, might have prevented this fatality.

ACCIDENT PREVENTION RECOMMENDATIONS

Toolbox Talks – Electrocution Hazards

What protective devices and procedures can you use to prevent electrocution?

What are the hazards?

What are the results?

What should we look for?

American Wire Gauge (AWG)

Cord Size	Handles Up To
#10 AWG	30 amps
#12 AWG	25 amps
#14 AWG	18 amps
#16 AWG	13 amps

Actual Incident: A 29-year- old male welder was electrocuted and died when he contacted an energized receptacle end of an extension cord. It was found that the welding unit and cord were incompatible; however, both the welding cord and extension cord were damaged allowing them to be used together. The result was an ungrounded system that killed a worker.

How do we prevent these results?

- ☐ ______________________
- ☐ ______________________
- ☐ ______________________
- ☐ ______________________
- ☐ ______________________
- ☐ ______________________

Let's talk about this site now.

ACCIDENT SUMMARY No. 13

Accident Type:	Collapse of Shoring
Weather Conditions:	Clear
Type of Operation:	Boring and Pipe Jacking Excavation
Size of Work Crew:	4
Collective Bargaining	Yes
Competent Safety Monitor on Site:	Yes
Safety and Health Program in Effect:	No
Was the Worksite Inspected Regularly:	Yes
Training and Education Provided:	Yes
Employee Job Title:	Pipe Welder
Age & Sex:	62-Male
Experience at this Type of Work:	18 years
Time on Project:	2½

BRIEF DESCRIPTION OF ACCIDENT

Four employees were boring a hole and pushing a 20-inch pipe casing under a road. The employees were in an excavation approximately 9 feet wide, 32 feet long and 7 feet deep. Steel plates 8' × 15' × ¾", being used as shoring, were placed vertically against the north and south walls of the excavation at approximately a 30 degree angle. There were no horizontal braces between the steel plates. The steel plate on the south wall tipped over, pinning an employee (who was killed) between the steel plate and the pipe casing. At the time the plate tipped over, a backhoe was being operated adjacent to the excavation.

INSPECTION RESULTS

As a result of its investigation, OSHA issued a citation for two alleged serious violations of its construction standards. OSHA's construction safety standards include several requirements which, if they had been followed here, might have prevented this fatality.

ACCIDENT PREVENTION RECOMMENDATIONS

____

____

____

____

____

____

Toolbox Talks – Caught-in or -Between

Actual Incident:
A 56-year-old truck driver was crushed when a crane tipped over and the crane's boom landed on the cab of the dump truck in which he was sitting. The crane had been lowering an empty four-yard concrete bucket, while booming out.

How do we prevent these results?

Let's talk about some site conditions:

How to prevent injury

- Ask for a nail gun with a sequential trigger mechanism.
- NEVER shoot towards yourself or a co-worker.
- Do not press the trigger unless the nose of the gun (contact element) is firmly pressed against the work material.
- NEVER walk around with your finger on the trigger.
- NEVER clean or clear jams or adjust a nail gun when it is connected to the air supply.
- Avoid nailing into knots and metal; nails are more likely to ricochet. Dense materials, like laminated beams, are also difficult to nail.
- NEVER remove or bypass safety devices, triggers, or contact springs.
- NEVER use a defective tool. If a tool is malfunctioning, it needs to be tagged and taken out of service.

CPWR

To read stories about nail gun injuries and see photos, visit
www.cpwr.com./nailguns

To learn more about CPWR, visit
www.cpwr.com.

For more safety and health information, visit
www.elcosh.org

 CPWR is the research, training, and service arm of the Building and Construction Trades Dept., AFL-CIO, and works to reduce or eliminate safety and health hazards construction workers face on the job. Production of this flier was supported by Grant OH008307 from the National Institute for Occupational Safety and Health (NIOSH). The contents are solely the responsibility of the authors and do not necessarily represent the official views of NIOSH.

HAZARD ALERT

Nail Guns

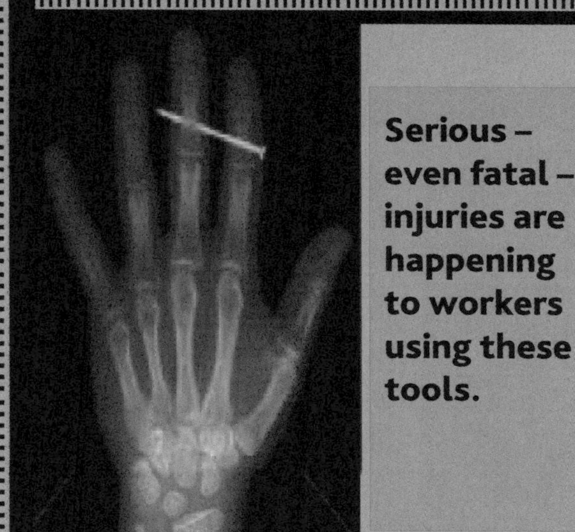

Serious – even fatal – injuries are happening to workers using these tools.

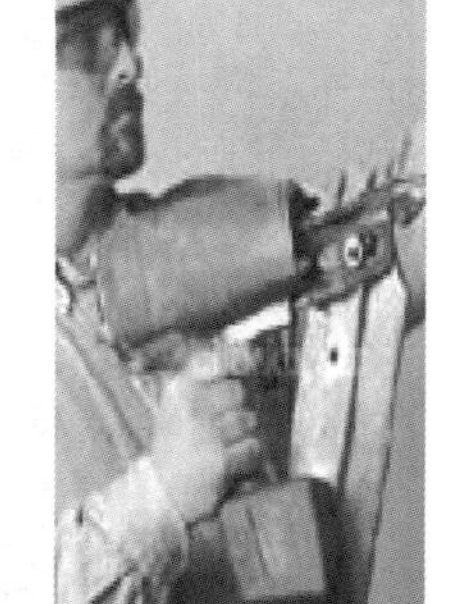
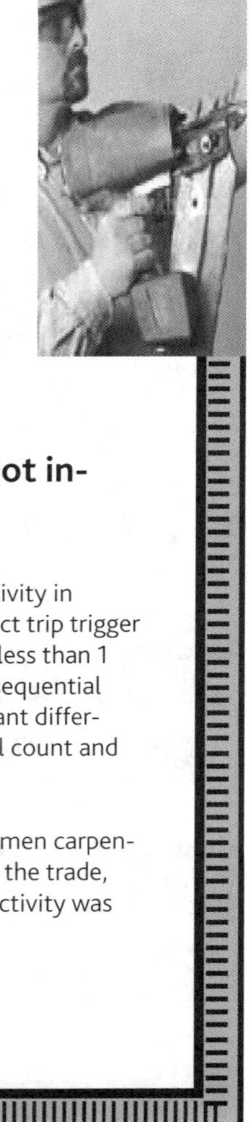

ESSENTIALS OF SAFETY

What's the problem?

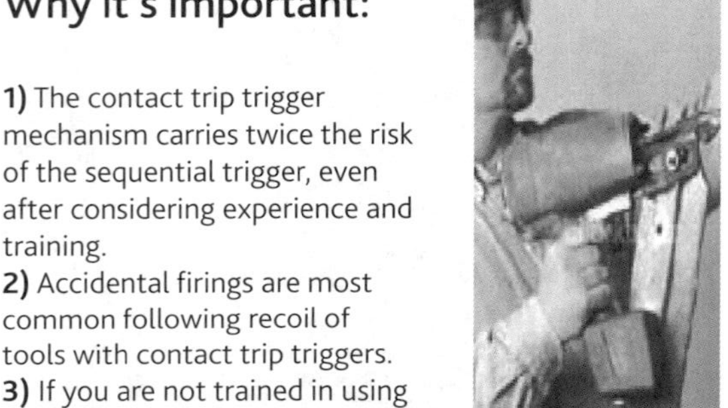

Nail guns are popular for a reason. They get the job done in a blink of an eye.

But that rapid-fire action can work against you. In a split second, a nail can enter your finger, your hand, or worse.

Nail gun injuries are much more common than people think. Most injuries involve puncture wounds to hands or fingers, but serious, even fatal, injuries are also associated with the use of these tools.

How most nail gun injuries happen

- Accidental or unintended firing, often associated with recoil of the tool after firing
- Ricocheting nails
- Nail going through work surface
- Airborne nails
- By-passed safety features
- Unsafe work practices
- Holding finger on contact trigger

Basic information about nail guns

Although there are many types of nail guns (framing, finishing, flooring, etc.),there are two common triggers:

Contact trip trigger mechanisms allow the tool to fire anytime the trigger and the nose of the gun (contact element) are both depressed. Trigger can be held down to allow bump or bounce nailing.

Sequential triggers require the nose of gun (contact element) to be depressed before the trigger is pulled. That avoids inadvertent discharge of nails.

WARNING:

The two triggers look exactly alike. You will not be able to tell the difference!

If you can "bump nail" by holding the trigger down, and bouncing the nose against a nailing surface, that is a contact trigger gun. Use extreme caution.

Why it's important:

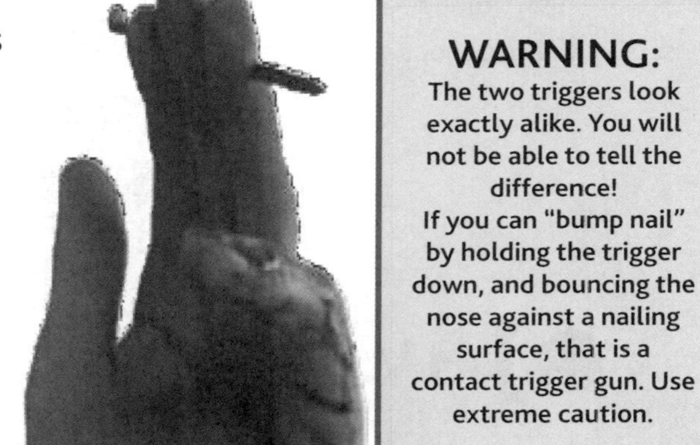

1) The contact trip trigger mechanism carries twice the risk of the sequential trigger, even after considering experience and training.
2) Accidental firings are most common following recoil of tools with contact trip triggers.
3) If you are not trained in using either of these tools, you are at high risk of injury.

"Faster" trigger does not increase productivity

A recent study measuring productivity in construction found that the contact trip trigger showed no significant difference (less than 1 percent) in productivity than the sequential trigger. Also, there was no significant difference between the two tools in nail count and placement.

The study, which involved journeymen carpenters with an average of 13 years in the trade, found that the difference in productivity was the worker, not the tool.

ACCIDENT SUMMARY No. 2

Accident Type:	Struck by Nail
Weather Conditions:	N/A
Type of Company:	General Contractors
Size of Work Crew:	17
Union or Non-union:	Union
Worksite Inspection?:	No
Designated Competent Person on Site?:	No
Employer Safety and Health Program?:	No
Training and Education for Employees?:	No
Craft of Deceased Employee(s):	Carpenter
Age;Sex	22; Male
Time of the Job:	3:00 p.m.
Time at the Task	Unknown

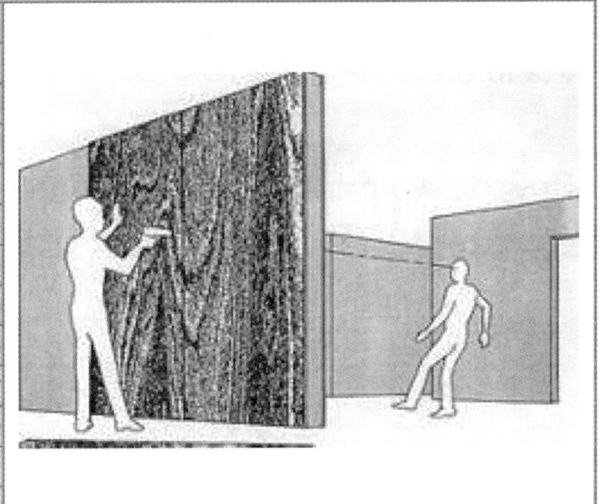

BRIEF DESCRIPTION OF ACCIDENT

A carpenter apprentice was killed when he was struck in the head by a nail that was fired from a powder actuated tool. The tool operator, while attempting to anchor a plywood form in preparation for pouring a concrete wall, fired the gun causing the nail to pass through the hollow wall. The nail travelled some twenty-seven feet before striking the victim. The tool operator had never received training in the proper use of the tool, and none of the employees in the area were wearing personal protective equipment.

INSPECTION RESULTS

Section not listed on original

ACCIDENT PREVENTION RECOMMENDATIONS

Toolbox Talks – Struck-by Hazards

Actual Incident:
A worker was cutting a six-inch steel water pipe with a gas-powered abrasive wheel when the saw kicked back and struck the worker's face shield, resulting in a laceration and two fractures to the nose.

How do we prevent these results?

Let's talk about some site conditions:

Appendix D to §1910.134

RESPIRATOR INFORMATION

MANCOMM®

Information for employees using respirators when not required under the standard

Respirators are an effective method of protection against designated hazards when properly selected and worn. Respirator use is encouraged, even when exposures are below the exposure limit, to provide an additional level of comfort and protection for workers. However, if a respirator is used improperly or not kept clean, the respirator itself can become a hazard to the worker. Sometimes, workers may wear respirators to avoid exposure to hazards, even if the amount of hazardous substance does not exceed the limits set by OSHA standards. If your employer provides respirators for your voluntary use, or if you provide your own respirator, you need to take certain precautions to be sure that the respirator itself does not present a hazard.

You should do the following:

1. Read and heed all instructions provided by the manufacturer on use, maintenance, cleaning, and care, and warnings regarding the respirator limitations.
2. Choose respirators certified for use to protect against the contaminant of concern. NIOSH, the National Institute for Occupational Safety and Health of the U.S. Department of Health and Human Services, certifies respirators. A label or statement of certification should appear on the respirator or respirator packaging. It will tell you what the respirator is designed for and how much it will protect you.
3. Do not wear your respirator into atmospheres containing contaminants for which your respirator is not designed to protect against. For example, a respirator designed to filter dust particles will not protect you against gases, vapors, or very small solid particles of fumes or smoke.
4. Keep track of your respirator so that you do not mistakenly use someone else's respirator.

I certify by signing this document that I have read and understood the above information.

______________________	______________________
Employee's Signature	Print Name
______________________	______________________
Date	Company Name

›Page 1-1

MEDICAL DETERMINATION FOR RESPIRATOR USE

MANCOMM®

____________________ Printed Employee Name

____ / ____ / ____ Date

____________________ Company Name

Respirator use for this person is: ☐ Approved ☐ Disapproved

Is a follow-up medical examination needed for this person? ☐ Yes ☐ No

List any limitations on respirator use related to the medical condition of this person or relating to the workplace conditions in which the respirator will be used:

____

____

____

Has this employee been provided with a copy of the Physician or Licensed Health Care Professional's (PLHCP) written recommendation regarding respirator use? ☐ Yes ☐ No

____________________ Signature (PLHCP)

____ / ____ / ____ Date

____________________ Company Name

____________________ Printed (PLHCP Name)

User Seal Checks

Each time you put on a tight-fitting respirator, you are required to perform a user seal check to make sure that an adequate seal is formed. If your face has changed in any way (for example, you've grown sideburns or a moustache, have lost a significant amount of weight, or have had dental changes), it is especially important to perform a seal check. Even a tiny gap between the respirator and your skin can let in contaminated air, causing injuries and possible death. An adequate seal means that the respirator will be able to do its job and keep you safe.

To perform a **positive pressure check**, do the following:

1. Close off the exhalation valve. For most respirators, you will have to first remove the exhalation valve cover.
2. Exhale gently into the facepiece.
3. The face fit is considered satisfactory if a slight positive pressure can be built up inside the facepiece without any evidence of outward leakage of air at the seal. Exhalation should cause a slight bulge. You should not be able to hear or feel any air leaking out.
4. Carefully replace the exhalation valve cover if you had to remove it to perform the test.

To perform a **negative pressure check**, do the following:

1. Close off the inlet opening of the canister or cartridge(s) by covering the opening with the palm of your hand or by replacing the filter seal(s). Some inlet openings of some cartridges cannot be covered with the palm of the hand. If this happens, cover the inlet opening of the cartridge with a thin latex or nitrile glove.
2. Inhale gently so that the facepiece collapses slightly.
3. Hold your breath for 10 seconds.
4. If the facepiece remains in its slightly collapsed condition and no inward leakage of air is detected, the tightness of the respirator is considered satisfactory.

Your respirator's manufacturer may have recommended other user seal check procedures. These recommended procedures may be used instead of a positive- or negative-pressure check so long as your employer has demonstrated that the manufacturer's procedures are equally effective.

If you find that the seal is not adequate on your respirator, **immediately inform your supervisor**. You will be given an opportunity to select a different respirator facepiece and be retested.

Remember, also, that a "fit test" is required before you use a respirator with a negative- or positive-pressure, tight-fitting facepiece. A fit test is a little more involved than a user seal check and is the subject of additional, in-depth training.

PPE for Workers Checklist

Protection	TYPICAL OPERATIONS OF CONCERN	YES	NO
EYE	Sawing, cutting, drilling, sanding, grinding, hammering, chopping, abrasive blasting, punch press operations, etc.		
	Pouring, mixing, painting, cleaning, siphoning, dip tank operations, dental and health care services, etc.		
	Battery charging, installing fiberglass insulation, compressed air or gas operations, etc.		
	Welding, cutting, laser operations, etc.		
FACE	Pouring, mixing, painting, cleaning, siphoning, dip tank operations, etc.		
	Welding, pouring molten metal, smithing, baking, cooking, drying, etc.		
	Cutting, sanding, grinding, hammering, chopping, pouring, mixing, painting, cleaning, siphoning, etc.		
HEAD	Work stations or traffic routes located under catwalks or conveyor belts, construction, trenching, utility work, etc.		
	Construction, confined space operations, building maintenance, etc.		
	Building maintenance; utility work; construction; wiring; work on or near communications, computer, or other high tech equipment; arc or resistance welding; etc.		
FEET	Construction, plumbing, smithing, building maintenance, trenching, utility work, grass cutting, etc.		
	Building maintenance; utility work; construction; wiring; work on or near communications, computer, or other high tech equipment; arc or resistance welding; etc.		
	Welding, foundry work, casting, smithing, etc.		
	Demolition, explosives manufacturing, grain milling, spray painting, abrasive blasting, work with highly flammable materials, etc.		
HANDS	Grinding, sanding, sawing, hammering, material handling, etc.		
	Pouring, mixing, painting, cleaning, siphoning, dip tank operations, health care and dental services, etc.		
	Welding, pouring molten metal, smithing, baking, cooking, drying, etc.		
	Building maintenance; utility work; construction; wiring; work on or near communications, computer, or other high tech equipment; arc or resistance welding; etc.		
BODY	Pouring, mixing, painting, cleaning, siphoning, dip tank operations, machining, sawing, battery charging, installing fiberglass insulation, compressed air or gas operations, etc.		
	Cutting, grinding, sanding, sawing, glazing, material handling, etc.		
	Welding, pouring molten metal, smithing, baking, cooking, drying, etc.		
	Pouring, mixing, painting, cleaning, siphoning, dip tank operations, etc.		
HEARING	Machining, grinding, sanding, work near conveyors, pneumatic equipment, generators, ventilation fans, motors, punch and brake presses, etc. Samples shown are: ear muffs (left) and earplugs (right)		

NOTE: Pictures of PPE are intended to provide a small sample of what the protection gear may look like. They are not to scale nor are they inclusive of all protection gear required and/or that is available.

Care, Maintenance, and Disposal of PPE

In order for it to do its job, personal protective equipment (PPE) must be properly cleaned, decontaminated, inspected, repaired (or replaced), and stored. Defective and broken equipment, used disposable PPE, and equipment that has exceeded the useful life suggested by the manufacturer must be disposed of properly.

There are two important general rules that should be followed regarding care and maintenance of all types of PPE:

- Always follow the manufacturer's recommendations for proper care and maintenance of PPE.
- Do not attempt repairs on PPE for which you are not qualified.

In addition, you should follow these guidelines for proper inspection, care, maintenance, and disposal of PPE:

- ***Eye and Face Protection:*** **Inspect** before each use, looking for cracks, pits, tears, scratches, that impair eyesight, loose elastic, and other defects. **Clean** with mild soap and water, making sure that the equipment is properly disinfected between uses if shared. **Store** in a dust-proof case in a clean, dry place. **Discard and replace** when defects, scratches, or pits impair eyesight or, if loose elastic interferes with the fit, replace the elastic or discard the equipment.
- ***Hearing Protection:*** **Inspect** before each use, looking for cracks, tears, dirt, imperfect seals or hard cushions on earmuffs, reusable earplugs that have become hard, and other defects. **Clean** reusable equipment with mild soap and water; never clean and reuse disposable earplugs. **Store** in a dust-proof case in a clean, dry place. **Discard** disposable earplugs after each use. **Repair or dispose** of reusable equipment when it no longer serves its function because of inadequate seals, improper fit, or other defects.
- ***Head Protection:*** **Inspect** before each use, looking for cracks, dents, holes, gouges, excessive wear, inadequate suspension systems, and tar, paint, oils, or other chemicals on the shell. **Clean** with mild soap and water, making sure that the equipment is properly disinfected between uses if shared. **Store** in a clean, dry place. **Discard and replace** when defects, holes, chemicals, etc. impair the protection the equipment offers. **Repair or replace** inadequate suspension systems. (Unless the manufacturer recommends otherwise, keep approximately 1¼ inches between your head and the shell.) **Dispose of and replace** any hard hat that has been subjected to any impact that may affect its integrity. (**Note:** Hard hats are designed to be protective for a single impact only; if a hard hat has taken a significant hit, it should be replaced.)
- ***Foot Protection:*** **Inspect** at least monthly, looking for excessive wear, tears, holes, broken stitching, broken laces, thin soles, loss of tread, slick soles, chemical damage, impairment to the steel toe, embedded materials in the sole, etc. **Clean** with mild soap or detergent and water. **Clean and oil** leather boots monthly. **Store** in a clean, dry place. **Repair or replace** immediately if a hole develops in the sole. **Repair or replace** foot protection with holes elsewhere if the holes impair in any way the protection offered by the footwear. **Dispose** of footwear contaminated with hazardous chemicals as if the footwear were hazardous waste.

Care, Maintenance, and Disposal of PPE (continued)

- ***Hand Protection:*** **Inspect** before each use, looking for correct fit, holes, cracks, tears, chemical damage, other damage, and excessive wear. **Clean** cloth gloves with mild soap or detergent and water. **Clean and treat** leather and thermal-protection gloves as recommended by the manufacturer. Do not clean disposable gloves; discard them after use. **Store** in a clean, dry place. **Discard and replace** gloves when holes or other damage appears or if they no longer fit. **Dispose** of gloves contaminated with hazardous chemicals as if they were hazardous waste.
- ***Respiratory Protection:*** **Inspect** before each use and during cleaning; inspect emergency respirators at least once a month and before and after each use. **Look** for wear and damage that could let in contaminated air, loose connections, deteriorating rubber parts, etc. Ensure that any required air or oxygen cylinders are fully charged. **Clean** according to manufacturer's recommendations, or by following this basic sequence:
  - Remove filters, cartridges, or canisters and disassemble the facepiece.
  - Wash components in warm water with mild detergent or cleaner. If the cleaner does not contain a disinfecting agent, then also soak the components for two minutes in a special disinfecting solution. Use a stiff bristle (not wire) brush to remove dirt.
  - Thoroughly rinse components in clean, warm, preferably running, water. Drain.
  - Hand dry components with a clean, lint-free cloth, or air dry.
  - Reassemble components.
  - Test for proper operation.

  Store in a clean, dry place out of direct sunlight and away from extreme temperatures and chemicals. Ensure that facepieces and valves are not being deformed. Emergency respirators must be stored where they will be accessible to the work area and in a compartment or cover that is clearly marked as containing an emergency respirator. **Dispose** of respirators by removing them from service and tagging them as defective.
- ***Body Protection and Protective Clothing:*** **Inspect** before each use (especially with chemical protective clothing), looking for poor fit, tears, holes, leaks, punctures, degradation, defects, and other damage. **Clean** and launder normal clothes as you usually would, making sure to properly decontaminate clothing contaminated with hazardous chemicals or other hazardous materials (such as asbestos) before washing. Do not clean disposable chemical protective clothing after use; instead, **discard** it. Properly decontaminate non-disposable chemical protective clothing before reuse. **Store** in a clean, dry place. **Dispose** of non-disposable chemical protective clothing whenever it has holes, tears, degradation, etc. **Discard** disposable chemical protective clothing after each use. **Discard** normal work clothes whenever there are holes or thin materials that expose skin to the elements.
- ***Personal Fall Protection Systems:*** **Inspect** before each use for knot and anchorage strength, mildew, wear, damage, deterioration, and defective components. **Clean and maintain** as instructed by the manufacturer. **Store** in clean, dry place out of direct sunlight and away from extreme temperatures and chemicals or as recommended by manufacturer. **Discard and replace** if defective or deteriorated. **Remove** system or components if subjected to impact loading until competent person determines it is not damaged and safe for use.

NIOSH HEAT STRESS FACTS

Protecting Yourself from Heat Stress

Heat stress, from exertion or hot environments, places workers at risk for illnesses such as heat stroke, heat exhaustion, or heat cramps.

Heat Stroke

A condition that occurs when the body becomes unable to control its temperature, and can cause death or permanent disability.

Symptoms

- High body temperature
- Confusion
- Loss of coordination
- Hot, dry skin or profuse sweating
- Throbbing headache
- Seizures, coma

First Aid

- Request immediate medical assistance.
- Move the worker to a cool, shaded area.
- Remove excess clothing and apply cool water to their body.

Heat Exhaustion

The body's response to an excessive loss of water and salt, usually through sweating.

Symptoms

- Rapid heart beat
- Heavy sweating
- Extreme weakness or fatigue
- Dizziness
- Nausea, vomiting
- Irritability
- Fast, shallow breathing
- Slightly elevated body temperature

First Aid

- Rest in a cool area.
- Drink plenty of water or other cool beverages.
- Take a cool shower, bath, or sponge bath.

Heat Cramps

Affect workers who sweat a lot during strenuous activity. Sweating depletes the body's salt and moisture levels.

Symptoms

- Muscle cramps, pain, or spasms in the abdomen, arms or legs

First Aid

- Stop all activity, and sit in a cool place.
- Drink clear juice or a sports beverage, or drink water with food.
  - Avoid salt tablets.
- Do not return to strenuous work for a few hours after the cramps subside.
- Seek medical attention if you have the following: heart problems, are on a low-sodium diet, or if the cramps do not subside within one hour.

Protect Yourself

Avoid heavy exertion, extreme heat, sun exposure, and high humidity when possible. When these cannot be avoided, take the following preventative steps:

- Monitor your physical condition and that of your coworkers for signs or symptoms of heat illnesses.
- Wear light-colored, loose-fitting, breathable clothing such as cotton.
  - Avoid non-breathable synthetic clothing.
- Gradually build up to heavy work.
- Schedule heavy work during the coolest parts of day.
- Take more breaks when doing heavier work, and in high heat and humidity.
  - Take breaks in the shade or a cool area.
- Drink water frequently. Drink enough water that you never become thirsty.
- Be aware that protective clothing or personal protective equipment may increase the risk of heat-related illnesses.

NIOSH 2010-114

Table 1: Specified Exposure Control Methods When Working with Materials Containing Crystalline Silica

Equipment / Task	Engineering and Work Practice Control Methods	Required Respiratory Protection and Minimum Assigned Protection Factor (APF)	
		≤ 4 hours/shift	> 4 hours/shift
(i) Stationary masonry saws	Use saw equipped with integrated water delivery system that continuously feeds water to the blade. Operate and maintain tool in accordance with manufacturer's instructions to minimize dust emissions.	None	None
(ii) Handheld power saws (any blade diameter)	Use saw equipped with integrated water delivery system that continuously feeds water to the blade. Operate and maintain tool in accordance with manufacturer's instructions to minimize dust emissions.		
	– When used outdoors.	None	APF 10
	– When used indoors or in an enclosed area.	APF 10	APF 10
(iii) Handheld power saws for cutting fiber-cement board (with blade diameter of 8 inches or less)	For tasks performed outdoors only: Use saw equipped with commercially available dust collection system. Operate and maintain tool in accordance with manufacturer's instructions to minimize dust emissions. Dust collector must provide the air flow recommended by the tool manufacturer, or greater, and have a filter with 99% or greater efficiency.	None	None
(iv) Walk-behind saws	Use saw equipped with integrated water delivery system that continuously feeds water to the blade. Operate and maintain tool in accordance with manufacturer's instructions to minimize dust emissions.		
	– When used outdoors.	None	None
	– When used indoors or in an enclosed area.	APF 10	APF 10
(v) Drivable saws	For tasks performed outdoors only: Use saw equipped with integrated water delivery system that continuously feeds water to the blade. Operate and maintain tool in accordance with manufacturer's instructions to minimize dust emissions.	None	None
(vi) Rig-mounted core saws or drills	Use tool equipped with integrated water delivery system that supplies water to cutting surface. Operate and maintain tool in accordance with manufacturer's instructions to minimize dust emissions..	None	None
(vii) Handheld and stand-mounted drills (including impact and rotary hammer drills)	Use drill equipped with commercially available shroud or cowling with dust collection system. Operate and maintain tool in accordance with manufacturer's instructions to minimize dust emissions. Dust collector must provide the air flow recommended by the tool manufacturer, or greater, and have a filter with 99% or greater efficiency and a filter-cleaning mechanism. Use a HEPA-filtered vacuum when cleaning holes.	None	None

Table 1 continued

Table 1: Specified Exposure Control Methods When Working with Materials Containing Crystalline Silica

Equipment / Task	Engineering and Work Practice Control Methods	Required Respiratory Protection and Minimum Assigned Protection Factor (APF)	
		≤ 4 hours/shift	> 4 hours/shift
(viii) Dowel drilling rigs for concrete	For tasks performed outdoors only: Use shroud around drill bit with a dust collection system. Dust collector must have a filter with 99% or greater efficiency and a filter-cleaning mechanism. Use a HEPA-filtered vacuum when cleaning holes.	APF 10	APF 10
(ix) Vehicle-mounted drilling rigs for rock and concrete	Use dust collection system with close capture hood or shroud around drill bit with a low-flow water spray to wet the dust at the discharge point from the dust collector.	None	None
	OR		
	Operate from within an enclosed cab and use water for dust suppression on drill bit.	None	None
(x) Jackhammers and handheld powered chipping tools	Use tool with water delivery system that supplies a continuous stream or spray of water at the point of impact.		
	– When used outdoors.	None	APF 10
	– When used indoors or in an enclosed area.	APF 10	APF 10
	OR		
	Use tool equipped with commercially available shroud and dust collection system. Operate and maintain tool in accordance with manufacturer's instructions to minimize dust emissions. Dust collector must provide the air flow recommended by the tool manufacturer, or greater, and have a filter with 99% or greater efficiency and a filter-cleaning mechanism.		
	– When used outdoors.	None	APF 10
	– When used	APF 10	APF 10
(xi) Handheld grinders for mortar removal (i.e., tuck-pointing)	Use grinder equipped with commercially available shroud and dust collection system. Operate and maintain tool in accordance with manufacturer's instructions to minimize dust emissions. Dust collector must provide 25 cubic feet per minute (cfm) or greater of airflow per inch of wheel diameter and have a filter with 99% or greater efficiency and a cyclonic pre-separator or filter-cleaning mechanism.	APF 10	APF 25

Table 1 continued

Table 1: Specified Exposure Control Methods When Working with Materials Containing Crystalline Silica

Equipment / Task	Engineering and Work Practice Control Methods	Required Respiratory Protection and Minimum Assigned Protection Factor (APF)	
		≤ 4 hours/shift	> 4 hours/shift
(xii) Handheld grinders for uses other than mortar removal	For tasks performed outdoors only: Use grinder equipped with integrated water delivery system that continuously feeds water to the grinding surface. Operate and maintain tool in accordance with manufacturer's instructions to minimize dust emissions.	None	None
	OR		
	Use grinder equipped with commercially available shroud and dust collection system. Operate and maintain tool in accordance with manufacturer's instructions to minimize dust emissions. Dust collector must provide 25 cubic feet per minute (cfm) or greater of airflow per inch of wheel diameter and have a filter with 99% or greater efficiency and a cyclonic pre-separator or filter-cleaning mechanism.		
	– When used outdoors.	None	None
	– When used indoors or in an enclosed area.	None	APF 10
(xiii) Walk-behind milling machines and floor grinders	Use machine equipped with integrated water delivery system that continuously feeds water to the cutting surface. Operate and maintain tool in accordance with manufacturer's instructions to minimize dust emissions.	None	None
	OR		
	Use machine equipped with dust collection system recommended by the manufacturer. Operate and maintain tool in accordance with manufacturer's instructions to minimize dust emissions. Dust collector must provide the air flow recommended by the manufacturer, or greater, and have a filter with 99% or greater efficiency and a filter-cleaning mechanism. When used indoors or in an enclosed area, use a HEPA-filtered vacuum to remove loose dust in between passes.	None	None
(xiv) Small drivable milling machines (less than half-lane)	Use a machine equipped with supplemental water sprays designed to suppress dust. Water must be combined with a surfactant. Operate and maintain machine to minimize dust emissions.	None	None

Table 1 continued

Table 1: Specified Exposure Control Methods When Working with Materials Containing Crystalline Silica

Equipment / Task	Engineering and Work Practice Control Methods	Required Respiratory Protection and Minimum Assigned Protection Factor (APF)	
		≤ 4 hours/shift	> 4 hours/shift
(xv) Large drivable milling machines (half-lane and larger)	For cuts of any depth on asphalt only: Use machine equipped with exhaust ventilation on drum enclosure and supplemental water sprays designed to suppress dust. Operate and maintain machine to minimize dust emissions.	None	None
	For cuts of four inches in depth or less on any substrate: Use machine equipped with exhaust ventilation on drum enclosure and supplemental water sprays designed to suppress dust. Operate and maintain machine to minimize dust emissions.	None	None
	OR Use a machine equipped with supplemental water spray designed to suppress dust. Water must be combined with a surfactant. Operate and maintain machine to minimize dust emissions.	None	None
(xvi) Crushing machines	Use equipment designed to deliver water spray or mist for dust suppression at crusher and other points where dust is generated (e.g., hoppers, conveyers, sieves/sizing or vibrating components, and discharge points). Operate and maintain machine in accordance with manufacturer's instructions to minimize dust emissions. Use a ventilated booth that provides fresh, climate-controlled air to the operator, or a remote control station.	None	None
(xvii) Heavy equipment and utility vehicles used to abrade or fracture silica-containing materials (e.g., hoe-ramming, rock ripping) or used during demolition activities involving silica-containing materials	Operate equipment from within an enclosed cab.	None	None
	When employees outside of the cab are engaged in the task, apply water and/or dust suppressants as necessary to minimize dust emissions.	None	None
(xviii) Heavy equipment and utility vehicles for tasks such as grading and excavating but not including: demolishing, abrading, or fracturing silica-containing materials	Apply water and/or dust suppressants as necessary to minimize dust emissions.	None	None
	OR When the equipment operator is the only employee engaged in the task, operate equipment from within an enclosed cab.	None	None

Table 1 end

Asbestos-Containing Materials

Note: The following list does not include every product or material that may contain asbestos. It is intended as a general guide to show which types of materials may contain asbestos. [source: *U.S. Environmental Protection Agency*]

Cement Pipes	Elevator Brake Shoes
Cement Wallboard	HVAC Duct Insulation
Cement Siding	Boiler Insulation
Asphalt Floor Tile	Breaching Insulation
Vinyl Floor Tile	Ductwork Flexible Fabric Connections
Vinyl Sheet Flooring	Cooling Towers
Flooring Backing	Pipe Insulation (corrugated air-cell, block, etc.)
Fire Doors	Heating and Electrical Ducts
Acoustical Plaster	Electrical Panel Partitions
Decorative Plaster	Electrical cloth
Textured Paints/Coatings	Electrical Wiring Insulation
Ceiling Tiles and Lay-In Panels	Chalkboards
Spray-Applied Insulation	Roofing Shingles
Blown-In Insulation	Roofing Felt
Fireproofing Materials	Base Flashing
Taping Compounds (thermal)	Thermal Paper Products
Packing Materials (for wall/floor penetrations)	Construction Mastics (floor tile, carpet, ceiling tile, etc.)
High-Temperature Gaskets	Caulking/Putties
Laboratory Hoods/Table Tops	Adhesives
Laboratory Gloves	Wallboard
Fire Blankets	Joint Compounds
Fire Curtains	Vinyl Wall Coverings
Elevator Equipment Panels	Spackling Compounds

Soil Conditions That Affect Rock / Soil Slope Stability

Condition	Requirement
Soil is fractured/unstable dry rock.	Downgrade to Type B.
Soil is fractured/unstable submerged rock.	Downgrade to Type C.
Soil is cemented (caliche, hardpan, etc.).	Classify as Type A.
Soil is fissured.	Downgrade from Type A to Type B.
Soil is subject to vibration.	Downgrade from Type A to Type B.
Soil has been previously disturbed.	Downgrade from Type A to Type B.
Soil is submerged or water is freely seeping through the sides of the excavation.	Downgrade from Type A to Type C. Downgrade from Type B to Type C.
Soil profile is layered with the layers dipping into the excavation on a slope of four horizontal to one vertical or steeper.	Downgrade from Type A to Type C. Downgrade from Type B to Type C.

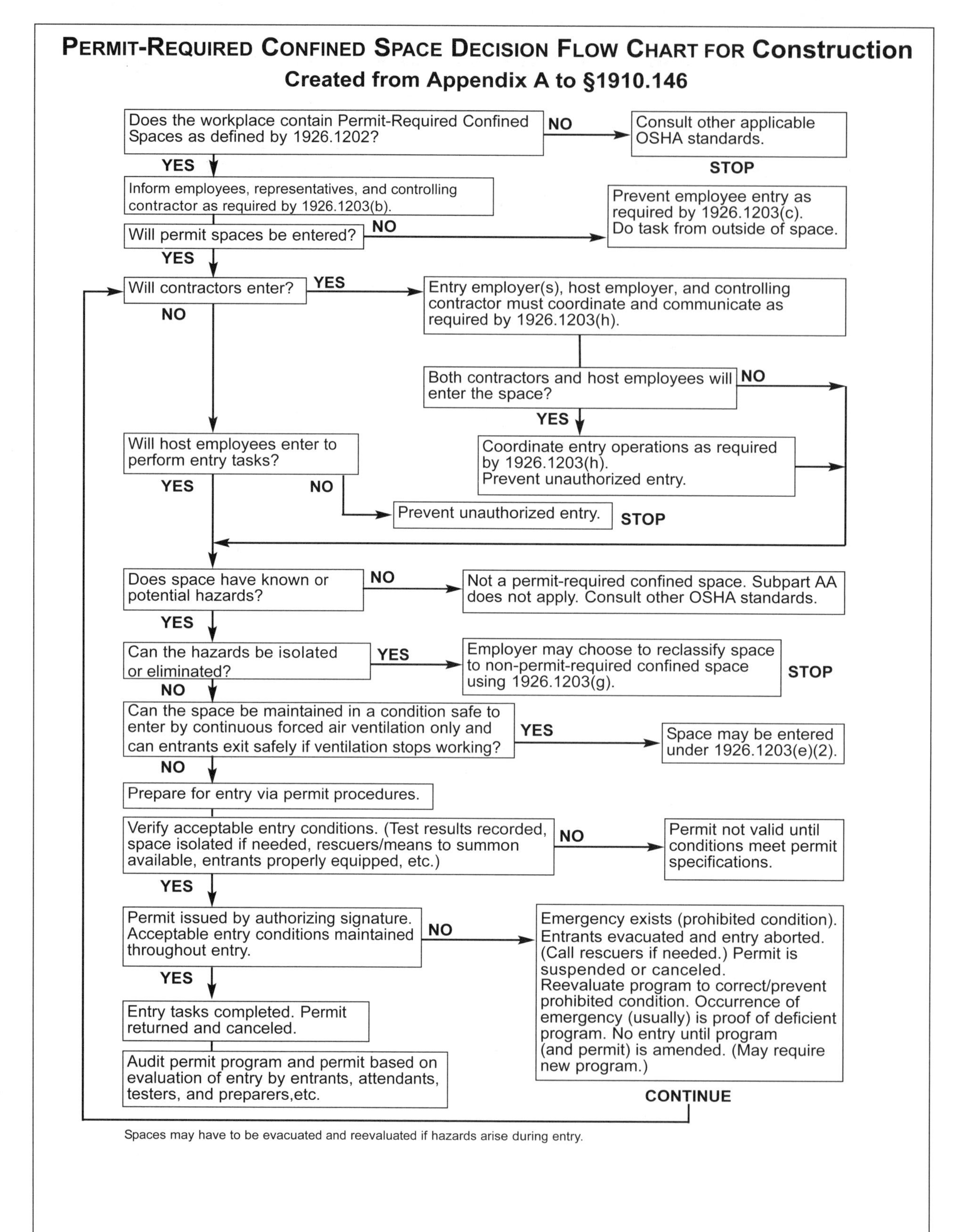
PERMIT-REQUIRED CONFINED SPACE DECISION FLOW CHART FOR Construction
Created from Appendix A to §1910.146
Does the workplace contain Permit-Required Confined Spaces as defined by 1926.1202?
NO
Consult other applicable OSHA standards.
STOP
YES
Inform employees, representatives, and controlling contractor as required by 1926.1203(b).
Will permit spaces be entered?
NO
Prevent employee entry as required by 1926.1203(c). Do task from outside of space.
YES
Will contractors enter?
YES
Entry employer(s), host employer, and controlling contractor must coordinate and communicate as required by 1926.1203(h).
NO
Both contractors and host employees will enter the space?
NO
YES
Coordinate entry operations as required by 1926.1203(h). Prevent unauthorized entry.
Will host employees enter to perform entry tasks?
YES
NO
Prevent unauthorized entry.
STOP
Does space have known or potential hazards?
NO
Not a permit-required confined space. Subpart AA does not apply. Consult other OSHA standards.
YES
Can the hazards be isolated or eliminated?
YES
Employer may choose to reclassify space to non-permit-required confined space using 1926.1203(g).
STOP
NO
Can the space be maintained in a condition safe to enter by continuous forced air ventilation only and can entrants exit safely if ventilation stops working?
YES
Space may be entered under 1926.1203(e)(2).
NO
Prepare for entry via permit procedures.
Verify acceptable entry conditions. (Test results recorded, space isolated if needed, rescuers/means to summon available, entrants properly equipped, etc.)
NO
Permit not valid until conditions meet permit specifications.
YES
Permit issued by authorizing signature. Acceptable entry conditions maintained throughout entry.
NO
Emergency exists (prohibited condition). Entrants evacuated and entry aborted. (Call rescuers if needed.) Permit is suspended or canceled. Reevaluate program to correct/prevent prohibited condition. Occurrence of emergency (usually) is proof of deficient program. No entry until program (and permit) is amended. (May require new program.)
YES
Entry tasks completed. Permit returned and canceled.
Audit permit program and permit based on evaluation of entry by entrants, attendants, testers, and preparers,etc.
CONTINUE
Spaces may have to be evacuated and reevaluated if hazards arise during entry.

Sample Confined Space Entry Permit as required by §1926.1206

Entry Permit: Pseudo Construction Co. Page 1 of 2

General Information

Identity (e.g., location) of the space: 1500 K Street, northwest corner.

Purpose of entry: Replace communication cable in sewer line.

Duration of entry: First entry on 10-15-15; complete work and terminate entry operations on 10-22-15.

Identify the physical hazards in the space:

(1) Engulfment—sewer water.

(2) Electrical—communication cables.

Describe the methods for isolating or controlling the physical hazards, or used to protect authorized entrants:

(1) Engulfment—Disconnect and lockout all sewer-system overflow pumps, and disconnect, blind, and lockout, all water lines within 100 feet of the work area, including lateral lines.

(2) Electrical—de-energize, tag, and ground all communication cables in work area. Use heavy duty, insulated work gloves for handling cables and conduits.

Identify the atmospheric hazards in the space (e.g., oxygen deficiency, flammable/explosive gases/vapors, others (including toxic particulates, gases, and vapors)):

(1) Oxygen deficiency—possibility that oxygen level may be less than 19.5%.

(2) Flammable/explosive gases/vapors—methane may be present.

(3) Other—hydrogen sulfide and/or carbon monoxide may be present.

Describe the methods for isolating or controlling the atmospheric hazards, or used to protect authorized entrants:

Space ventilation—mechanical ventilation systems set at 100% outside air. Where possible, open additional manholes to increase air circulation. Use portable blowers to augment natural ventilation if needed. Repeat atmospheric testing after ventilating the space for 10 minutes.

Describe the determination made to show that if the ventilation system stops working, atmospheric hazards will remain at safe levels long enough for entrants to recognize the problem and safely exit the space:

Fully ventilated the unoccupied space and got non-detect readings for methane and hydrogen sulfide. After securing the mechanical ventilation system, ports, and portable blowers, found that both methane and hydrogen sulfide reached maximum limits in two hours, which is well outside the 60-minute period used to monitor and record atmospheric-hazard values in the space.

Planned Conditions

Safe conditions and/or monitoring determined for physical hazards:

(1) Water levels—pooling water level inside PRCS not to exceed 2 inches; survey PRCS every hour to assess pooling water level and seeping water from piping and sewer.

(2) Electrical—visually confirm every hour that electrical cables remain disconnected and tagged.

(3) Early-warning system—when alarm sounds, evacuate authorized employees immediately.

Safe levels of atmospheric hazards:

Hazard	Minimum limit	Maximum limit
Oxygen	19.50%	23.00%
Flammable gas / vapors (specify):		
(1) Methane	0% LFL	10% LFL
Others (specify):		
(1) Hydrogen sulfide	0 ppm	20 ppm
(2) Carbon Monoxide	0 ppm	35 ppm

Atmospheric-testing/-monitoring results: (see attachment)

Sample Confined Space Entry Permit as required by §1926.1206

Entry Permit: Pseudo Construction Co. Page 2 of 2

Current Entry Supervisor:

Name: S. Smith Signature/initial: /s/
Name: J. Jones Signature/initial: /s/
Name: ________________ Signature/initial: ____________________
Name: ________________ Signature/initial: ____________________
Name: ________________ Signature/initial: ____________________

Current Attendants:

Name: R. Montero Name: ________________
Name: K. Chang Name: ________________
Name: ________________ Name: ________________
Name: ________________ Name: ________________
Name: ________________ Name: ________________

Authorized entrants: Refer to daily tracking roster for names of authorized entrants.

Identity of the rescue service: (Check one)

_X_ *Non-entry rescue*. Identity of the non-entry rescue service: the attendant
___ *Entry rescue*. Identity of the entry rescue service:

Identity of the emergency service: Washington, DC Fire and Rescue Service, Station 1. Each attendant and entry supervisor has cell phone pre-programmed with DC Fire and Rescue Service emergency hotline number (e.g., 555-123-4567).

Name and signature/initials of the entry supervisor who first verified this entry permit and authorized initial entry into this PRCS:
Name: ________________ Signature/initial: ____________________

Equipment

Methods of communication between attendants and authorized entrants: Type X walkie-talkies.
Equipment Needed:
(1) Lighting equipment: two explosion-proof and waterproof lights with 50-ft. extension cords.
(2) Ventilation: one explosion-proof blower and 15-foot of duct.
(3) Atmospheric monitoring: One multi-gas meter equipped with sensors for oxygen, methane, hydrogen sulfide, and carbon monoxide.
(4) Heating: N/A
(5) Guardrails, barriers, and covers: one portable guardrail and tent assembly.
(6) Controlled descent/retrieval systems: N/A
(7) Ladders: a 16-foot fixed ladder for access/egress.
(8) Scaffolding: N/A
(9) Early warning system: remote high-water-level detector and alarm installed 100 yards upstream from the work area in the inflow conduit.
(10) Rescue equipment: one retrieval system with adjustable harness and 50-foot of retrievable lanyard.
(11) Other: portable gas-powered electric generator with GFCI.

Other Information

Information not documented elsewhere on this permit (see § 1926.1206(o)):
(1) **Describe any condition making it unsafe to remove an entrance cover, and how the condition was eliminated**: Determined that the entrance cover was vented, which eliminated the potential for hazardous pressure conditions to exist in the space.
(2) **Describe any actions taken to guard holes and openings into the space from falling individuals and objects**: A portable guardrail system was erected around the entry point, in addition to placing warning cones to divert pedestrian traffic around the space.
(3) **Describe the method used for entering and exiting the space**: a 16-foot fixed ladder.
Additional information: Hot-work permit issued for welding work.

Entry Permit Cancellation

Reason for cancellation: Completed work in PRCS.
Name and signature/initials of the individual who cancelled the entry permit:
Name: J. Jones Signature/initial: /s/
Date and time this entry permit was cancelled: 10-22-15; 4:30 PM

Confined Space Safe Entry Checklist

Use the following checklist, which is based on those produced by the National Institute for Occupational Safety and Health and the North Carolina Department of Labor's Occupational Safety and Health Division, to evaluate a confined space before you enter it. **Do not enter or allow anyone else to enter a confined space until you have considered every question and have determined that the space is safe! Note**: This checklist is based on the requirements for general industry; however it is also useful for the construction industry.

	YES	NO
Is entry necessary?	❑	❑
TESTING		
Are the instruments used in atmospheric testing properly calibrated?	❑	❑
Was the atmosphere in the confined space tested?	❑	❑
Was oxygen at least 19.5% and not more than 23.5%?	❑	❑
Were toxic, flammable, or oxygen-displacing gases or vapors present?	❑	❑
_____ Hydrogen sulfide		
_____ Carbon monoxide		
_____ Methane		
_____ Carbon dioxide		
_____ Other *(list)* ________________________		
Have all authorized entrants or their authorized representatives been provided an opportunity to observe the periodic testing of the permit space?	❑	❑
MONITORING		
Will the atmosphere in the space be monitored while work is going on?	❑	❑
Will the atmosphere in the space be monitored continuously while work is going on?	❑	❑
Will the atmosphere in the space be monitored periodically while work is going on? If so, at what interval? ________________________	❑	❑

Remember, atmospheric changes can occur due to the work procedure, the product stored, or outside influences. The atmosphere may be safe when you enter, but it can change very quickly!

Confined Space Safe Entry Checklist (continued)

CLEANING	YES	NO
Has the space been cleaned before entry is made?	❑	❑
Was the space steamed?	❑	❑
If the space was steamed, was it allowed to cool?	❑	❑
VENTILATION		
Are the appropriate operator's instructions for the ventilation system available?	❑	❑
Has the space been ventilated before entry?	❑	❑
Will ventilation be continued during entry?	❑	❑
Is the air intake for the ventilation system located in an area that is free of combustible dusts and vapors and toxic substances?	❑	❑
If the atmosphere was found unacceptable and then ventilated, was it retested before entry?	❑	❑
ISOLATION		
Does the company have a lock-out/tag-out program?	❑	❑
Have you been trained under the lock-out/tag-out program?	❑	❑
Are the specific lock-out/tag-out procedures available at the worksite?	❑	❑
Has the space been isolated from other systems?	❑	❑
Has electrical equipment been locked out?	❑	❑
Have disconnects been used where possible?	❑	❑
Has mechanical equipment been blocked, chocked, and disengaged where necessary?	❑	❑
Have lines under pressure been blanked and bled?	❑	❑

Confined Space Safe Entry Checklist (continued)

Clothing and Equipment	**Yes**	**No**
Is special clothing (e.g., boots, chemical suits, glasses, etc.) required?	❑	❑
If so, specify: ______________________________		
Is special equipment required (e.g., rescue equipment, communications equipment, etc.)? If so, specify: ______________________________	❑	❑
Are special tools required?	❑	❑
If so, specify: ______________________________		
Have all entrants been trained and certified in the use of any required personal protective equipment?	❑	❑
Respiratory Protection		
Does the employer have a respiratory protection program?	❑	❑
Are the affected employees included in this program?	❑	❑
Are MSHA/NIOSH-approved respirators of the type required available at the worksite?	❑	❑
Is respiratory protection required? If so, specify type (e.g., air-purifying, supplied air, self-contained breathing apparatus, etc.): ______________________________	❑	❑
Can you get through the opening into the confined space with a respirator on? (If you don't know, find out before you try to enter.)	❑	❑
Training		
Have you been trained in the proper use of a respirator?	❑	❑
Have you received first-aid and cardiopulmonary (CPR) training?	❑	❑
Have you been trained in confined space entry and do you know what to look for?	❑	❑

Confined Space Safe Entry Checklist (continued)

STANDBY/RESCUE	YES	NO
Will there be a standby person (i.e., attendant) on the outside in constant visual or auditory communication with the person(s) on the inside?	❑	❑
Will the standby person be able to see and/or hear the person(s) inside at all times?	❑	❑
Has the standby person been trained in rescue?	❑	❑
Will safety lines and harnesses be required to remove a person?	❑	❑
Are company rescue procedures available to be followed in the event of an emergency?	❑	❑
Are you familiar with emergency rescue procedures?	❑	❑
Do you know who to notify and how to notify them in the event of an emergency?	❑	❑
Is the evacuation alarm in proper working order?	❑	❑
PERMIT		
Has a confined space entry permit been issued?	❑	❑
Have you reviewed the permit, making sure that the space has been tested by a qualified person and that the space is safe for entry?	❑	❑
Have you reviewed the permit so that you know what hazards may be present, all required precautions and equipment, and what work is to be done?	❑	❑
Does the permit include a list of emergency telephone numbers?	❑	❑

Filter Shade Numbers for Protection Against Radiant Energy (§1926.102(c), Table E-1)

Welding operation	Shade number
Shielded metal-arc welding 1/16-, 3/32-, 1/8-, 5/32-inch diameter electrodes	10
Gas-shielded arc welding (nonferrous) 1/16-, 3/32-, 1/8-, 5/32-inch diameter electrodes	11
Gas-shielded arc welding (ferrous) 1/16-, 3/32-, 1/8-, 5/32-inch diameter electrodes	12
Shielded metal-arc welding 3/16-, 7/32-, 1/4-inch diameter electrodes	12
5/16-, 3/8-inch diameter electrodes	14
Atomic hydrogen welding	10 – 14
Carbon-arc welding	14
Soldering	2
Torch brazing	3 or 4
Light cutting, up to 1 inch	3 or 4
Medium cutting, 1 inch to 6 inches	4 or 5
Heavy cutting, over 6 inches	5 or 6
Gas welding (light), up to 1/8-inch	4 or 5
Gas welding (medium), 1/8-inch to 1/2-inch	5 or 6
Gas welding (heavy), over 1/2-inch	6 or 8

Different Forklifts = Different Hazards

The different types of forklifts (reproduced from the Cal/OSHA regulations at 8 CCR §3664) and the major risks posed by them are listed below:

- IT-2
  - Tipping over (high risk), especially at excessive speeds

FIGURE IT-2

- HIGH LIFT TRUCK
- COUNTERBALANCED TRUCK
- CANTILEVER TRUCK
- RIDER TRUCK
- FORK LIFT TRUCK

- IT-3
  - Tipping over (very high risk) because of no counterbalance, especially with top-heavy loads
  - Being hit by falling objects

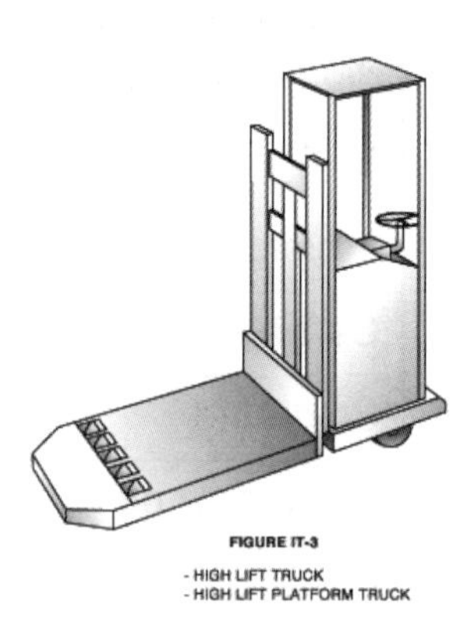

FIGURE IT-3

- HIGH LIFT TRUCK
- HIGH LIFT PLATFORM TRUCK

- IT-4
  - Striking objects or running over feet because of obstructed view
  - Being hit by falling objects

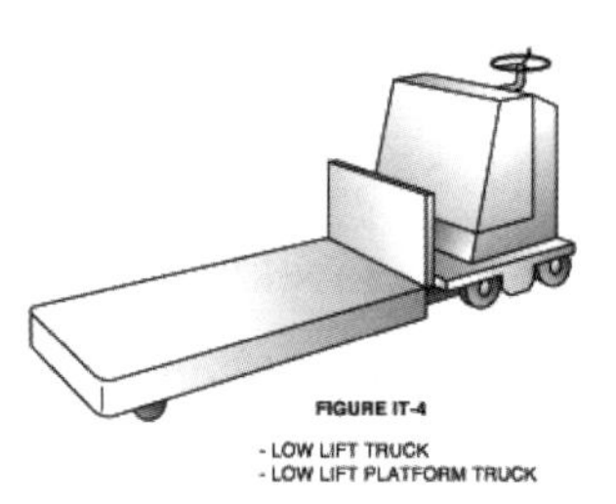

FIGURE IT-4

- LOW LIFT TRUCK
- LOW LIFT PLATFORM TRUCK

- IT-5
  - Striking objects or running over feet because of obstructed view
  - Being hit by falling objects

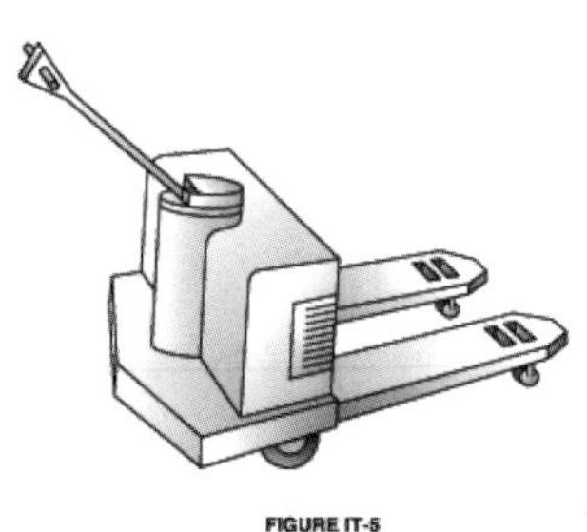

FIGURE IT-5

- MOTORIZED HAND TRUCK
- PALLET TRUCK

ESSENTIALS OF SAFETY

Different Forklifts = Different Hazards (continued)

- IT-6
  - Striking objects or running over feet because of obstructed view
  - Being hit by falling objects

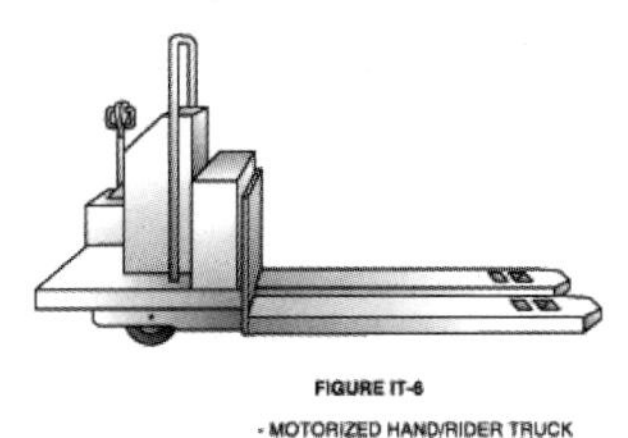

FIGURE IT-6
- MOTORIZED HAND/RIDER TRUCK

- IT-7
  - Tipping over from heavy or unbalanced loads

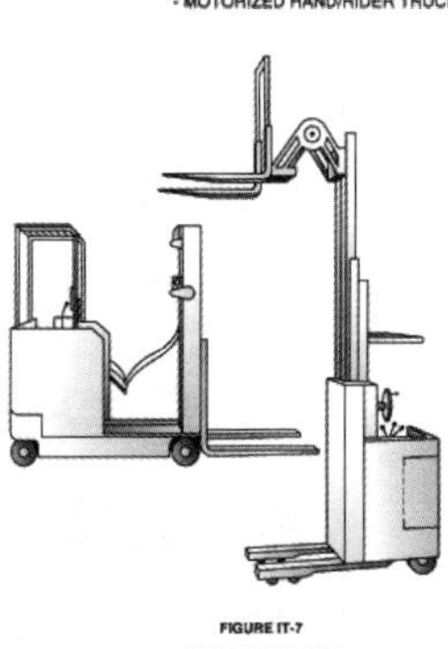

FIGURE IT-7
- REACH RIDER TRUCKS

- IT-8
  - Striking objects because of obstructed view or poor clearance

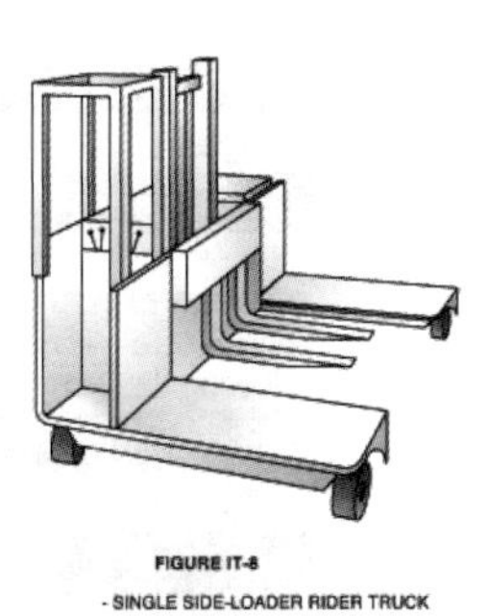

FIGURE IT-8
- SINGLE SIDE-LOADER RIDER TRUCK

- IT-9
  - Falling off (use belt or guardrails for fall protection)
  - Striking overhead objects

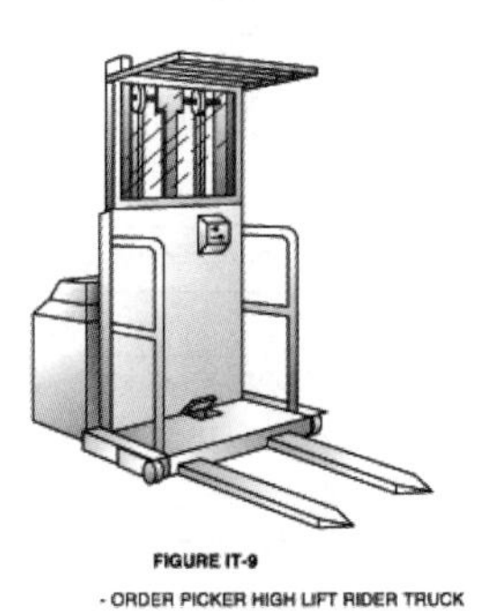

FIGURE IT-9
- ORDER PICKER HIGH LIFT RIDER TRUCK

- IT-10
  - Tipping over
  - Falling off (use belt or guardrails for fall protection)
  - Striking overhead objects
  - Losing control on rough surface (due to narrow track)

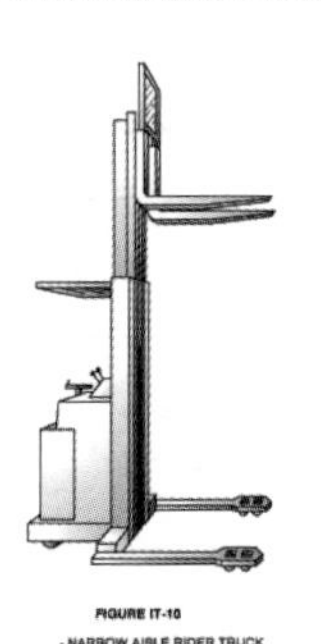

FIGURE IT-10
- NARROW AISLE RIDER TRUCK
- STRADDLE TRUCK

FORKLIFT OPERATOR'S DAILY CHECKLIST

INSTRUCTIONS

Fill in top half of box for pre-shift inspection. (If applicable)

Fill in bottom half of box for post-shift inspection. (If applicable)

Write notes at bottom of sheet

Vehicle Number: ____________________ **Meter:** __________ (Beginning of week) **Date:** ___/___/_____ (Beginning of week)

	MONDAY			TUESDAY			WEDNESDAY			THURSDAY			FRIDAY			SATURDAY			SUNDAY		
	SHIFT			SHIFT			SHIFT			SHIFT			SHIFT			SHIFT			SHIFT		
	1	2	3	1	2	3	1	2	3	1	2	3	1	2	3	1	2	3	1	2	3
General Condition																					
Tires																					
Fuel Level																					
Safety Equipment																					
Fluid Levels																					
Hoses																					
Belts																					
Cables																					
Screens																					
Horns and Alarms																					
Battery Charge																					
Control Levers																					
Gauges																					
Comm. System																					
Brakes																					
Controls																					
Steering System																					
Lift Controls																					
Forks																					
Mast																					
Operator Initials	_	_	_	_	_	_	_	_	_	_	_	_	_	_	_	_	_	_	_	_	_

Notes: (Please include date, shift and initials)

YOU ARE HEREBY AUTHORIZED TO MAKE COPIES FOR INTERNAL USE ONLY

ESSENTIALS OF SAFETY

HazCom – List of Hazardous Chemicals Workshop

	Date	Product Identifier (as Referenced on SDS)	Manufacturer	Where Used (Department)	Emergency Telephone Number
1					
2					
3					
4					
5					
6					
7					
8					
9					
10					
11					
12					
13					
14					
15					
16					
17					
18					
19					
20					

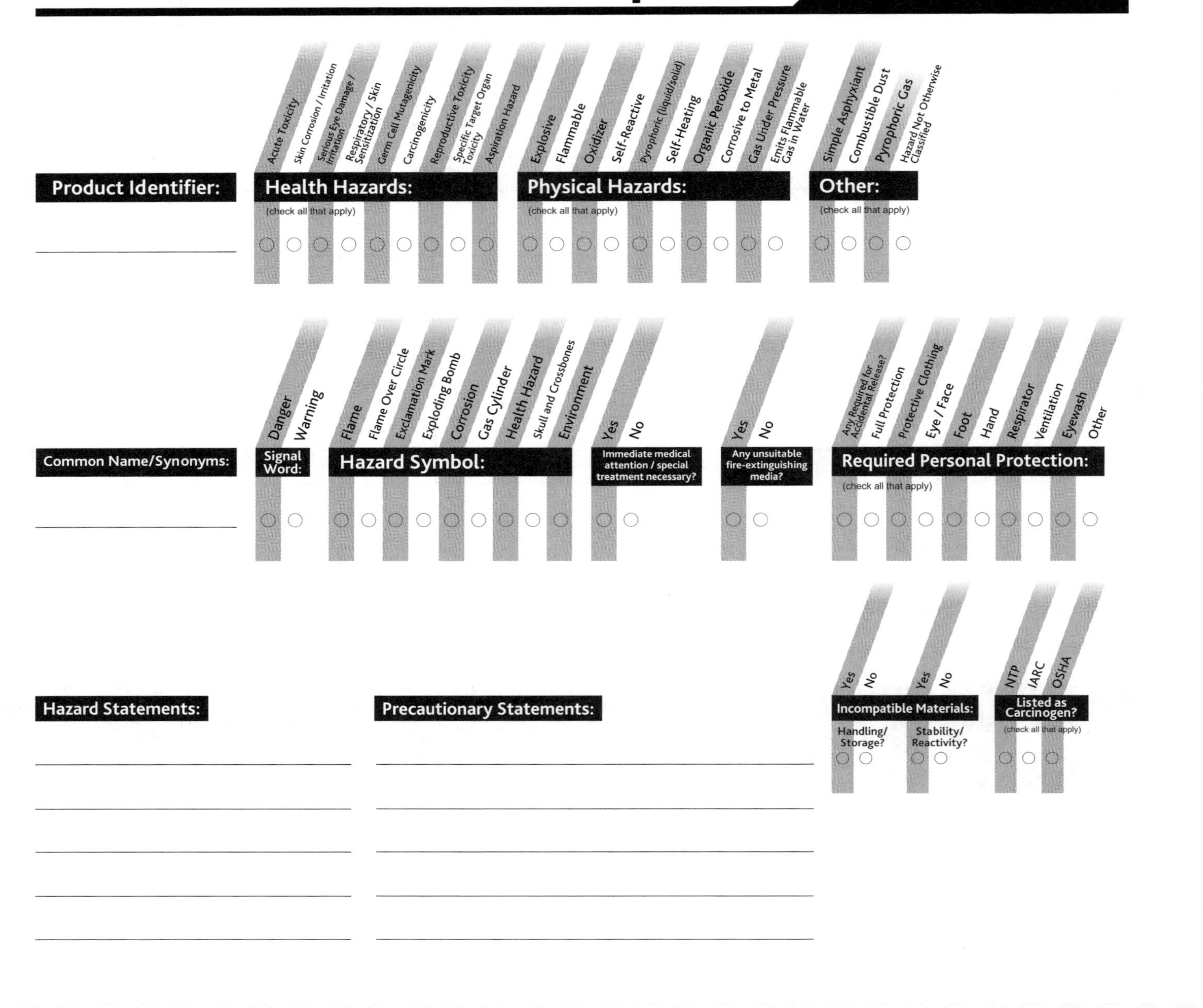

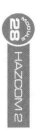

ESSENTIALS OF SAFETY

HazCom - SDS Workshop

Product Identifier:

Health Hazards: (check all that apply)
- Acute Toxicity
- Skin Corrosion / Irritation
- Serious Eye Damage / Irritation
- Respiratory / Skin Sensitization
- Germ Cell Mutagenicity
- Carcinogenicity
- Reproductive Toxicity
- Specific Target Organ Toxicity
- Aspiration Hazard

Physical Hazards: (check all that apply)
- Explosive
- Flammable
- Oxidizer
- Self-Reactive
- Pyrophoric (liquid/solid)
- Self-Heating
- Organic Peroxide
- Corrosive to Metal
- Gas Under Pressure
- Emits Flammable Gas in Water

Other: (check all that apply)
- Simple Asphyxiant
- Combustible Dust
- Pyrophoric Gas
- Hazard Not Otherwise Classified

Common Name/Synonyms:

Signal Word:
- Danger
- Warning

Hazard Symbol:
- Flame
- Flame Over Circle
- Exclamation Mark
- Exploding Bomb
- Corrosion
- Gas Cylinder
- Health Hazard
- Skull and Crossbones
- Environment

Immediate medical attention / special treatment necessary?
- Yes
- No

Any unsuitable fire-extinguishing media?
- Yes
- No

Required Personal Protection: (check all that apply)
- Any Required for Accidental Release?
- Full Protection
- Protective Clothing
- Eye / Face
- Foot
- Hand
- Respirator
- Ventilation
- Eyewash
- Other

Hazard Statements:

Precautionary Statements:

Incompatible Materials:
- Handling/ Storage? Yes / No
- Stability/ Reactivity? Yes / No

Listed as Carcinogen? (check all that apply)
- NTP
- IARC
- OSHA

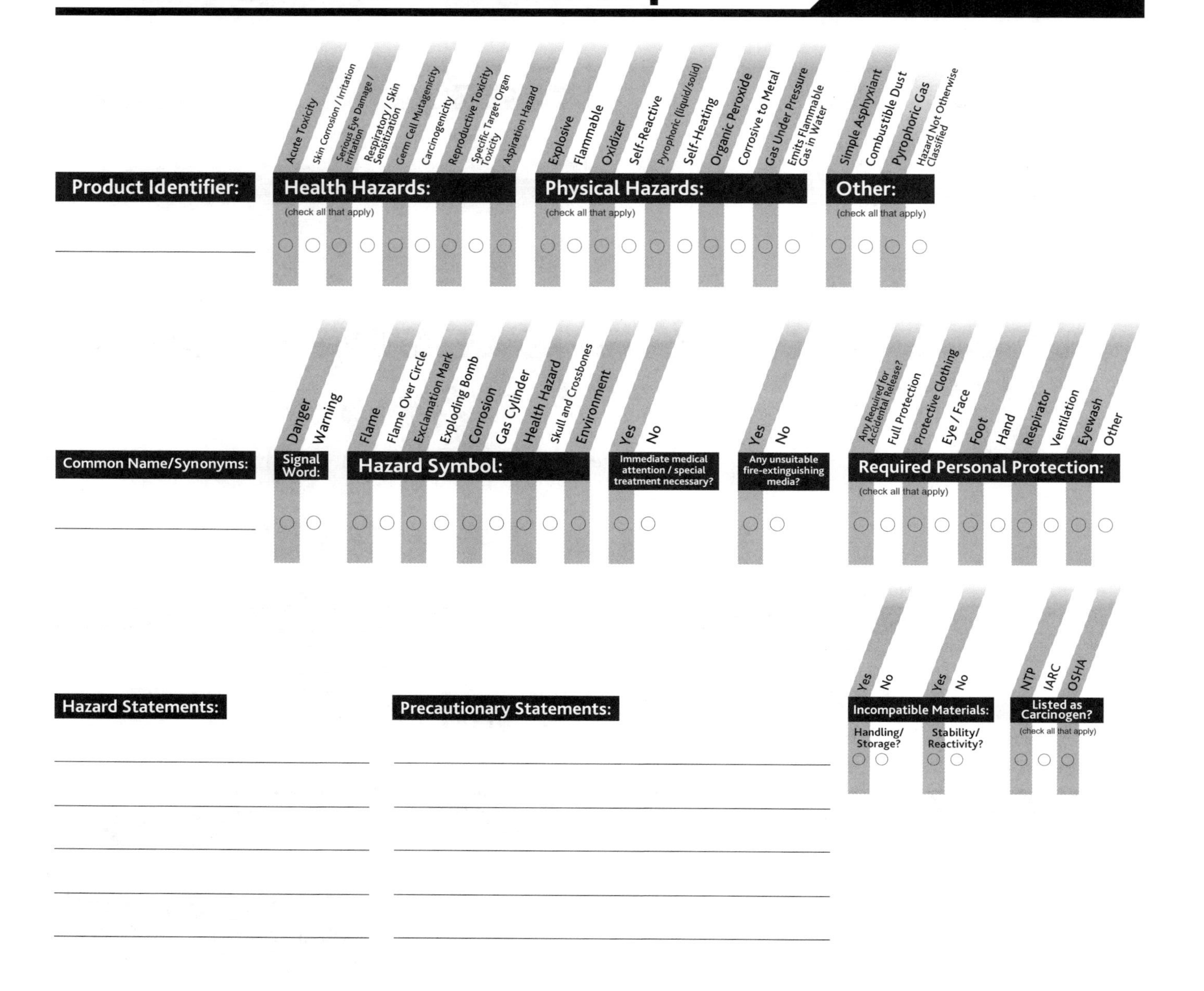

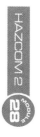

HazCom - SDS Workshop

Product Identifier:

____

Health Hazards: (check all that apply)
- ○ Acute Toxicity
- ○ Skin Corrosion / Irritation
- ○ Serious Eye Damage / Irritation
- ○ Respiratory / Skin Sensitization
- ○ Germ Cell Mutagenicity
- ○ Carcinogenicity
- ○ Reproductive Toxicity
- ○ Specific Target Organ Toxicity
- ○ Aspiration Hazard

Physical Hazards: (check all that apply)
- ○ Explosive
- ○ Flammable
- ○ Oxidizer
- ○ Self-Reactive
- ○ Pyrophoric (liquid/solid)
- ○ Self-Heating
- ○ Organic Peroxide
- ○ Corrosive to Metal
- ○ Gas Under Pressure
- ○ Emits Flammable Gas in Water

Other: (check all that apply)
- ○ Simple Asphyxiant
- ○ Combustible Dust
- ○ Pyrophoric Gas
- ○ Hazard Not Otherwise Classified

Common Name/Synonyms:

____

Signal Word:
- ○ Danger
- ○ Warning

Hazard Symbol:
- ○ Flame
- ○ Flame Over Circle
- ○ Exclamation Mark
- ○ Exploding Bomb
- ○ Corrosion
- ○ Gas Cylinder
- ○ Health Hazard
- ○ Skull and Crossbones
- ○ Environment

Immediate medical attention / special treatment necessary?
- ○ Yes
- ○ No

Any unsuitable fire-extinguishing media?
- ○ Yes
- ○ No

Required Personal Protection: (check all that apply)
- ○ Any Required for Accidental Release?
- ○ Full Protection
- ○ Protective Clothing
- ○ Eye / Face
- ○ Foot
- ○ Hand
- ○ Respirator
- ○ Ventilation
- ○ Eyewash
- ○ Other

Incompatible Materials:

Handling/ Storage?
- ○ Yes
- ○ No

Stability/ Reactivity?
- ○ Yes
- ○ No

Listed as Carcinogen? (check all that apply)
- ○ NTP
- ○ IARC
- ○ OSHA

Hazard Statements:

____

____

____

____

____

Precautionary Statements:

____

____

____

____

____

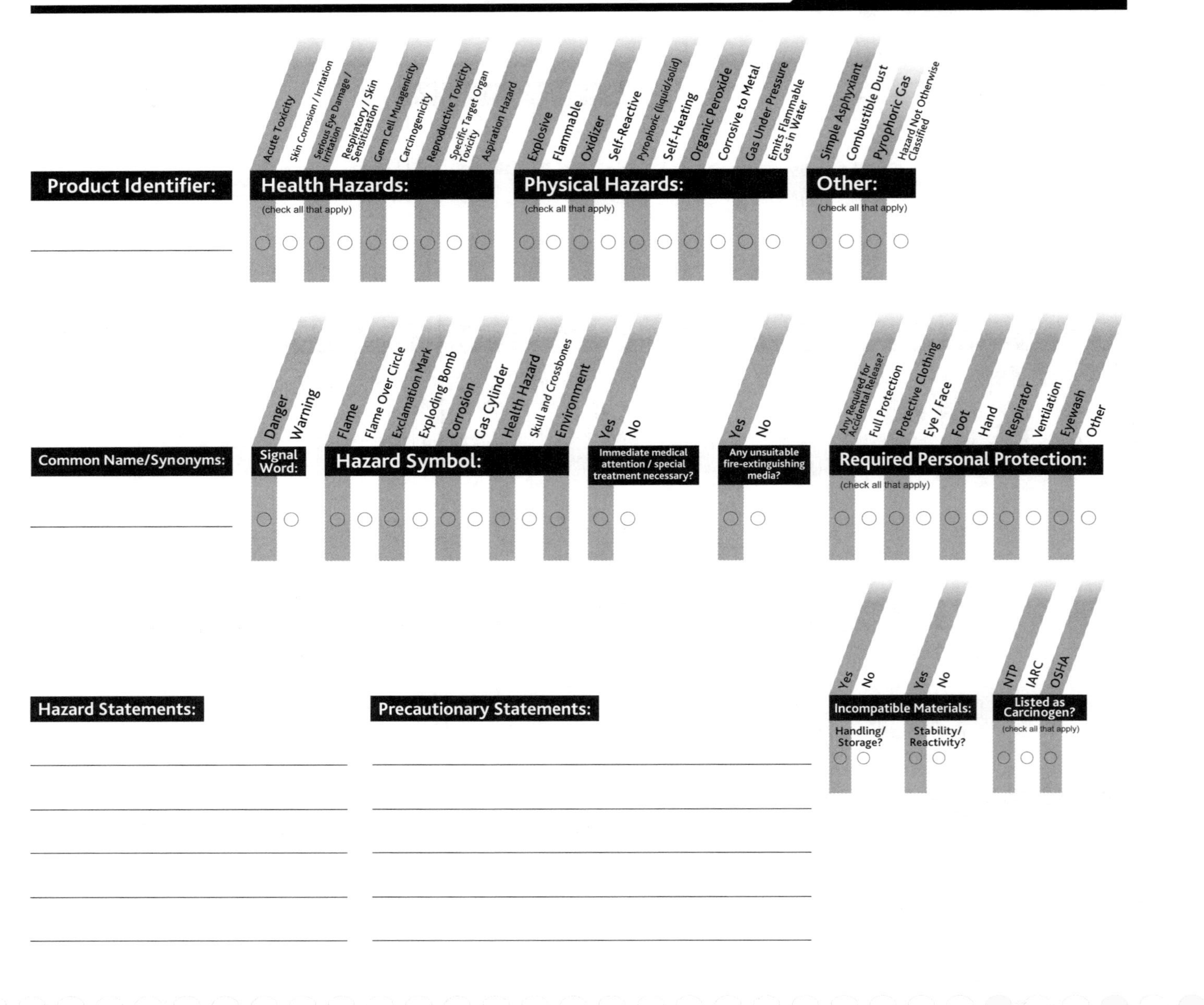

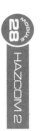

ESSENTIALS OF SAFETY

HazCom - SDS Workshop

Product Identifier:

____

Health Hazards: (check all that apply)

- ○ Acute Toxicity
- ○ Skin Corrosion / Irritation
- ○ Serious Eye Damage / Irritation
- ○ Respiratory / Skin Sensitization
- ○ Germ Cell Mutagenicity
- ○ Carcinogenicity
- ○ Reproductive Toxicity
- ○ Specific Target Organ Toxicity
- ○ Aspiration Hazard

Physical Hazards: (check all that apply)

- ○ Explosive
- ○ Flammable
- ○ Oxidizer
- ○ Self-Reactive
- ○ Pyrophoric (liquid/solid)
- ○ Self-Heating
- ○ Organic Peroxide
- ○ Corrosive to Metal
- ○ Gas Under Pressure
- ○ Emits Flammable Gas in Water

Other: (check all that apply)

- ○ Simple Asphyxiant
- ○ Combustible Dust
- ○ Pyrophoric Gas
- ○ Hazard Not Otherwise Classified

Common Name/Synonyms:

____

Signal Word:

- ○ Danger
- ○ Warning

Hazard Symbol:

- ○ Flame
- ○ Flame Over Circle
- ○ Exclamation Mark
- ○ Exploding Bomb
- ○ Corrosion
- ○ Gas Cylinder
- ○ Health Hazard
- ○ Skull and Crossbones
- ○ Environment

Immediate medical attention / special treatment necessary?

- ○ Yes
- ○ No

Any unsuitable fire-extinguishing media?

- ○ Yes
- ○ No

Required Personal Protection: (check all that apply)

- ○ Any Required for Accidental Release?
- ○ Full Protection
- ○ Protective Clothing
- ○ Eye / Face
- ○ Foot
- ○ Hand
- ○ Respirator
- ○ Ventilation
- ○ Eyewash
- ○ Other

Hazard Statements:

____

____

____

____

____

Precautionary Statements:

____

____

____

____

____

Incompatible Materials:

Handling/ Storage?

- ○ Yes
- ○ No

Stability/ Reactivity?

- ○ Yes
- ○ No

Listed as Carcinogen? (check all that apply)

- ○ NTP
- ○ IARC
- ○ OSHA

HazCom - Labeling Workshop

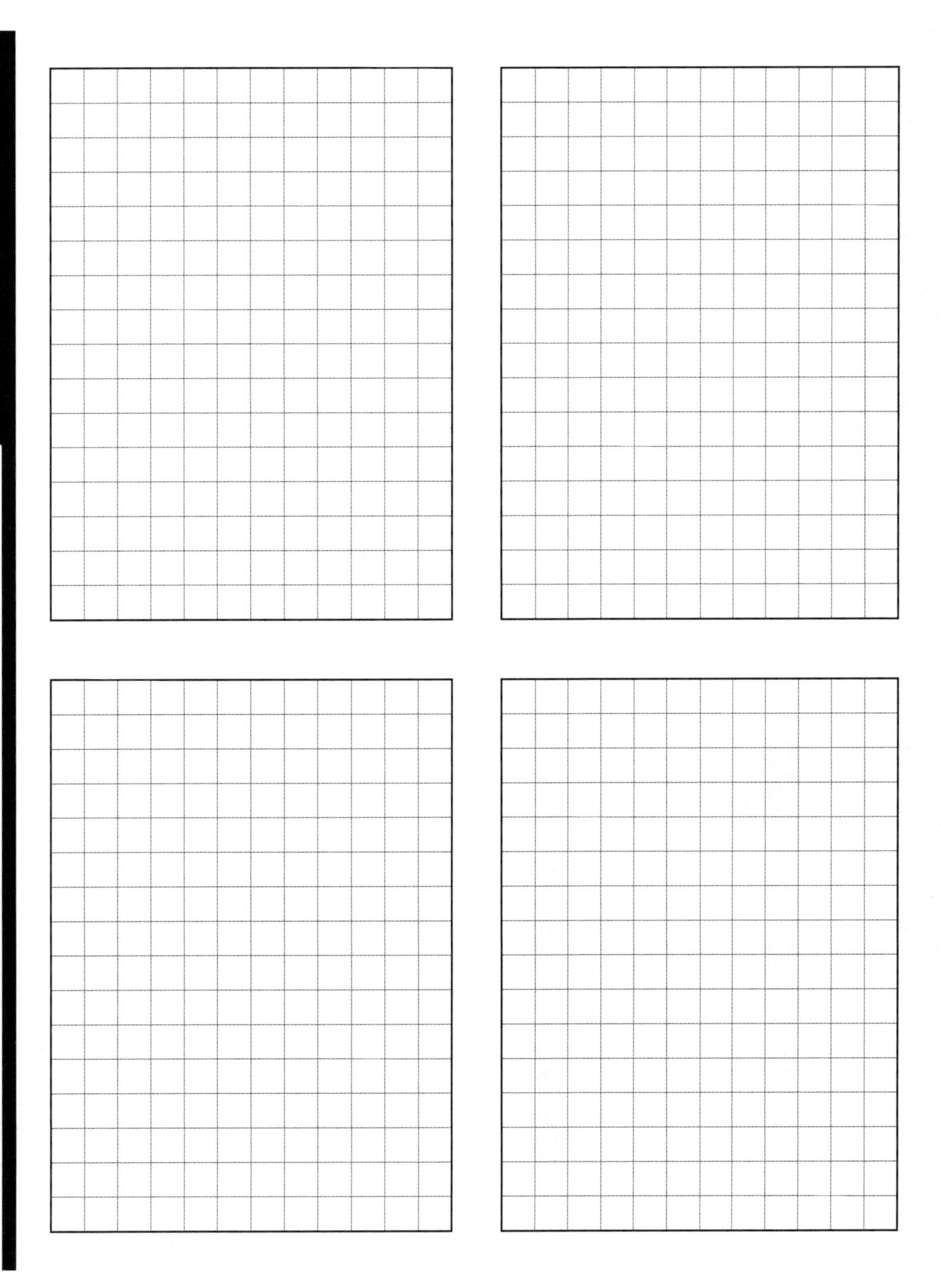

ESSENTIALS OF
SAFETY

HazCom - Training Workshop

Name of Chemical: ____________________

Methods of Detection:

Visual appearance: ____________________

Odor of the chemical: ____________________

Other: ____________________

Hazards:

Physical Hazard: ____________________

Health Hazard: ____________________

Other Hazard: ____________________

Health effects:

Eye: ____________________

Skin: ____________________

Ingestion: ____________________

Inhalation: ____________________

As part of the training, explain the labeling system and review the SDS.

Methods of Protection:

Engineering Controls: ____________________

Safe Work Practices: ____________________

Personal Protective Equipment: ____________________

Storage Procedures: ____________________

Clean-up Procedures: ____________________

First-Aid Procedures: ____________________

Fire-Fighting Procedures: ____________________

Other Protections: ____________________

HazCom Sample Checklist

This is a sample checklist to which additional items should be added or deleted depending on your needs.

	General Requirements:
	Is there a written hazard communication program that deals with labels, safety data sheets (SDSs), and employee training?
	Does the program include a list of the hazardous chemicals known to be in the workplace?
	Does the program discuss the methods that will be used to inform employees of the hazards of non-routine tasks and the hazards associated with chemicals contained in unlabeled pipes in employee work areas?
	Does the program discuss the methods that will be used to provide other employers in multi-employer workplaces onsite access to SDSs, information on precautionary measures for both normal operating conditions and foreseeable emergencies, and the employer's workplace labeling system?
	Is each container of hazardous chemicals properly labeled according to the requirements of the HazCom standard?
	Is there an SDS readily available for each hazardous chemical in the workplace?
	Is there an employee hazard communication training program?
	Does the program identify a person(s) responsible for ensuring proper labeling, maintaining SDSs, and conducting training?
	Is the written program made available to employees upon request?
	Are employees trained on the hazardous chemicals in their work area at the time of their initial assignment and whenever a new chemical hazard is introduced into the workplace?
	Does the Information Provided to Employees Include:
	The requirements of the HazCom standard?
	Any operations in their work area where hazardous chemicals are present?
	The location and availability of the written hazard communication program (including the required list of hazardous chemicals) and SDSs?
	Does Employee Training Cover:
	Methods and observations to detect the presence or release of hazardous chemicals in the work area?
	The hazards of the chemicals in the work area?
	Measures employees can take to protect themselves (e.g., appropriate work practices, emergency procedures, PPE, etc.)?
	An explanation of labels received on shipped containers?
	An explanation of the employer's workplace labeling system?
	An explanation of SDSs, including the order of information and how to obtain and use the appropriate hazard information?

NINE EASY THINGS TO GET YOUR PROGRAM STARTED

If these recommended practices appear challenging, here are some simple steps you can take to get started. Completing these steps will give you a solid base from which to take on some of the more structured actions presented in the recommended practices.

1. ALWAYS SET SAFETY AND HEALTH AS THE TOP PRIORITY

Tell your workers that making sure they finish the day and go home safely is the way you do business. Assure them that you will work with them to find and fix any hazards that could injure them or make them sick.

2. LEAD BY EXAMPLE

Practice safe behaviors yourself and make safety part of your daily conversations with workers.

3. IMPLEMENT A REPORTING SYSTEM

Develop and communicate a simple procedure for workers to report any injuries, illnesses, incidents (including near misses/close calls), hazards, or safety and health concerns without fear of retaliation. Include an option for reporting hazards or concerns anonymously.

4. PROVIDE TRAINING

Train workers on how to identify and control hazards using, for example, OSHA's Hazard Identification Training Tool.

5. CONDUCT INSPECTIONS

Inspect the job site with workers and ask them to identify any activity, piece of equipment, or material that concerns them. Use checklists and other resources, such as OSHA's Construction Industry Digest, to help identify problems.

6. COLLECT HAZARD CONTROL IDEAS

Talk with workers about ideas on safety improvements throughout the project.

7. IMPLEMENT HAZARD CONTROLS

Assign workers the task of choosing, implementing, and evaluating the solutions.

8. ADDRESS EMERGENCIES

Identify foreseeable emergency scenarios and develop instructions on what to do in each case. Meet to discuss these procedures and post them in a visible location at the job site.

9. MAKE IMPROVEMENTS

Set aside a regular time to discuss safety and health issues, with the goal of identifying ways to improve the program.

https://www.osha.gov/shpguidelines/docs/8524_OSHA_Construction_Guidelines_R4.pdf

CORE ELEMENTS OF THE RECOMMENDED PRACTICES FOR SAFETY AND HEALTH PROGRAMS IN CONSTRUCTION

MANAGEMENT LEADERSHIP	• Top management demonstrates its commitment to eliminating hazards and to continuously improving workplace safety and health, communicates that commitment to workers, and sets program expectations and responsibilities. • Managers at all levels make safety and health a core organizational value, establish safety and health goals and objectives, provide adequate resources and support for the program, and set a good example.
WORKER PARTICIPATION	• Workers and their representatives are involved in all aspects of the program—including setting goals, identifying and reporting hazards, investigating incidents, and tracking progress. • All workers, including contractors and temporary workers, understand their roles and responsibilities under the program and what they need to do to effectively carry them out. • Workers are encouraged and have means to communicate openly with management and to report safety and health concerns or suggest improvements, without fear of retaliation. • Any potential barriers or obstacles to worker participation in the program (for example, language, lack of information, or disincentives) are removed or addressed.
HAZARD IDENTIFICATION AND ASSESSMENT	• Procedures are put in place to continually identify workplace hazards and evaluate risks. • Safety and health hazards from routine, nonroutine, and emergency situations are identified and assessed. • An initial assessmen t of existing hazards, exposures, and control measures is followed by periodic inspections and reassessments, to identify new hazards. • Any incidents are investigated with the goal of identifying the root causes. • Identified hazards are prioritized for control.
HAZARD PREVENTION AND CONTROL	• Employers and workers cooperate to identify and select methods for eliminating, preventing, or controlling workplace hazards. • Controls are selected according to a hierarchy that uses engineering solutions first, followed by safe work practices, administrative controls, and finally personal protective equipment (PPE). • A plan is developed tha t ensures controls are implemented, interim protection is provided, progress is tracked, and the effectiveness of controls is verified.
EDUCATION AND TRAINING	• All workers are trained to understand how the program works and how to carry out the responsibilities assigned to them under the program. • Employers, managers, and supervisors receive training on safety concepts and their responsibility for protecting workers' rights and responding to workers' reports and concerns. • All workers are trained to recognize workplace hazards and to understand the control measures that have been implemented.
PROGRAM EVALUATION AND IMPROVEMENT	• Control measures are periodically evaluated for effectiveness. • Processes are established to monitor program performance, verify program implementation, and identify program shortcomings and opportunities for improvement. • Necessary actions are taken to improve the program and overall safety and health performance.
COMMUNICATION AND COORDINATION FOR EMPLOYERS ON MULTIEMPLOYER WORKSITES	• General contractors, contractors, and staffing agencies commit to providing the same level of safety and health protection to all employees. • General contractors, contractors, subcontractors, and staffing agencies commmunicate the hazards present at the worksite and the hazards that work of contract workers may create on site. • General contractors establish specifications and qualifications for contractors and staffing agencies. • Prior to beginning work, general contractors, contractors, and staffing agencies coordinate on work planning and scheduling to identify and resolve any conflicts that could impact safety or health.

https://www.osha.gov/shpguidelines/docs/8524_OSHA_Construction_Guidelines_R4.pdf

Safety and Health Program Audit Tool

Company/Site: ______________________ Address: ______________________

____

Auditor Name: ______________________ Auditor Phone #: ______________________

Auditor Signature: ______________________ Date: ______________________

Section 1: Management Leadership

Action Item	Not Implemented	Partially Implemented	Implemented with only Minor Deficiencies	Fully Implemented	Evidence of Implementation
Management implements and communicates a written policy supporting the safety and health program.					
Management defines specific goals and expectations for the program, and plans to achieve the goals.					
Management allocates appropriate resources (funds and time) to accomplish goals and manage the program.					
Management assigns responsibility and accountability for implementing and maintaining the program.					
All workers know who has been assigned responsibility for the program.					
Management integrates safety and health into planning and budgeting processes.					
Management recognizes worker contributions to workplace safety and health.					
Management routinely demonstrates visible commitment to the program.					

Section 2: Worker Participation

Action Item	Not Implemented	Partially Implemented	Implemented with only Minor Deficiencies	Fully Implemented	Evidence of Implementation
Workers are encouraged to participate in the program, have the means to participate, and feel comfortable participating and providing input on safety and health issues.					
Workers are assigned roles or are otherwise involved in all aspects of the program.					
Workers have access to information they need to understand safety and health hazards and control measures in the workplace.					
Workers are able to participate without encountering language, skill, or education barriers; restrictions on participating during work time; or fear of retaliation or discrimination.					

https://www.osha.gov/shpguidelines/docs/SHP_Audit_Tool.pdf

Safety and Health Program Audit Tool (continued)

Section 2: Worker Participation (continued)

Action Item	Not Implemented	Partially Implemented	Implemented with only Minor Deficiencies	Fully Implemented	Evidence of Implementation
Workers know how to report an injury, illness, hazard, or concern, including good catches/near misses.					
Workers consistently report injuries, illnesses, hazards, and concerns, including good catches/near misses.					
Reports of injuries, illnesses, hazards, or other concerns are acknowledged promptly.					
Reports of injuries, illnesses, hazards, or other concerns are resolved promptly, after seeking worker input.					

Section 3: Hazard Identification and Assessment

Action Item	Not Implemented	Partially Implemented	Implemented with only Minor Deficiencies	Fully Implemented	Evidence of Implementation
Written materials such as injury logs, safety data sheets, medical reports, workplace inspection results, incident investigation reports, and manufacturers' literature are reviewed to help identify hazards.					
The workplace is inspected regularly to identify conditions that pose or could pose a safety concern. Inspections cover all areas and activities and include plant and transportation vehicles.					
The workplace is evaluated to identify worker exposure to health hazards.					
Incidents (including close calls/near misses) are investigated to identify any hazards previously unrecognized or inadequately controlled. Investigations focus on identifying the root cause(s) of each incident.					
Hazards associated with emergencies and non-routine operations are identified.					
Interim controls are adopted while permanent controls are being determined.					
All identified hazards are characterized with respect to the severity of potential outcomes, likelihood an event or exposure will occur, and number of workers who might be exposed.					
All serious and recognized hazard are addressed immediately, while remaining hazards are prioritized for further control.					

SAFETY AND HEALTH PROGRAM AUDIT TOOL (CONTINUED)

Section 4: Hazard Prevention and Control

Action Item	Not Implemented	Partially Implemented	Implemented with only Minor Deficiencies	Fully Implemented	Evidence of Implementation
Options for controlling hazards are identified using sources such as OSHA, NIOSH, industry best practices, and input from workers.					
Controls are selected according to the "hierarchy of controls", emphasizing in order of priority: elimination, substitution, engineering controls, administrative controls, and PPE.					
A hazard control plan is used to plan and prioritize controls, and track and verify their installation.					
Responsibility for installing or implementing controls is assigned to persons with power or ability to implement the controls.					
Interim controls are used when permanent controls cannot be immediately implemented.					
Controls are in place to protect workers during emergencies and nonroutine operations.					
Once installed, controls are monitored to ensure workers understand their use and application and to verify they are effective.					
Controls are inspected and maintained.					

Section 5: Education and Training

Action Item	Not Implemented	Partially Implemented	Implemented with only Minor Deficiencies	Fully Implemented	Evidence of Implementation
Managers, supervisors, and workers understand the elements of the safety and health program and how to participate in it.					
Employers, managers, and supervisors understand: fundamental concepts of hazard identification and control; procedures for responding to workers' reports of injuries, illnesses, and incidents; incident investigation techniques; their responsibilities under the OSH Act; and workers' rights guaranteed under the Act.					
Workers understand the employers' responsibilities under the program.					
Each worker understands his or her own role in the program.					
Workers know who to contact with concerns or questions, and understand the procedures for reporting injuries, incidents, hazards, and concerns.					

Safety and Health Program Audit Tool (continued)

Section 5: Education and Training (continued)

Action Item	Not Implemented	Partially Implemented	Implemented with only Minor Deficiencies	Fully Implemented	Evidence of Implementation
Workers know they have a right to participate in the program and report injuries without fear of retaliation or discrimination.					
Workers can ask questions, receive answers, and provide feedback during and after training.					
Employers, managers, and supervisors understand their responsibilities under the OSH Act; procedures for responding to workers' reports of injury, illness, or concern; techniques for identifying and controlling hazards; and fundamentals of incident investigation.					
Supplemental training is provided when a change in the workplace could introduce new or increased hazards.					
Supplemental training is provided when a worker is assigned a new task or given a new assignment.					
Training is provided in a language and at a literacy level that all workers can understand.					

Section 6: Program Evaluation and Improvement

Action Item	Not Implemented	Partially Implemented	Implemented with only Minor Deficiencies	Fully Implemented	Evidence of Implementation
Performance indicators are used to track progress towards program goals.					
Performance is tracked using both lagging and leading indicators.					
Performance data is analyzed and shared with workers.					
An initial review and subsequent annual reviews evaluate the program to ensure it is fully implemented and functioning as planned.					
Workers are involved in all program review activities.					
The program is modified as needed to correct shortcomings.					

Section 7: Communication and Coordination for Host Employers, Contractors, and Staffing Agencies

Action Item	Not Implemented	Partially Implemented	Implemented with only Minor Deficiencies	Fully Implemented	Evidence of Implementation
Before coming onsite, the host employer and any contractors or staffing agencies determine which among them will implement and maintain the various parts of the safety and health program.					

Safety and Health Program Audit Tool (continued)

Section 7: Communication and Coordination for Host Employers, Contractors, and Staffing Agencies (continued)

Action Item	Not Implemented	Partially Implemented	Implemented with only Minor Deficiencies	Fully Implemented	Evidence of Implementation
Before coming onsite all contractors, staffing agencies, and their workers are informed of the hazards that may be present, the controls in place to address the hazards, and who to contact to report an injury, illness, or concern.					
Before coming onsite, the host employer provides contractors and staffing agencies the opportunity to conduct site visits or inspections and to review injury and illness records and other safety and health information.					
Before coming onsite, contractors or staffing agencies inform the host employer of any hazards arising from their work onsite, the controls in place to address the hazards, and who to contact if they have a safety concern.					
Host employers communicate with contractors and staffing agencies to determine which will implement and maintain the various parts of the safety and health program.					
A mechanism is established to exchange information about hazards present in the workplace and measures in place to prevent or control them for all contractors to use to report injuries, hazards, and concerns.					
Host employers include any safety-related specifications and qualifications requirements for contractors and staffing agencies in contracts and bid documents.					
Host employers coordinate with contractors and staffing agencies to ensure work is planned and scheduled to minimize impacts on safety.					
Temporary workers are adequately trained and equipped before arriving at the worksite.					
Safety and health policies and procedures of host employers, contractors, and staffing agencies are consistent and understood by all workers onsite.					
Host employers and staffing agencies ensure enough trained and equipped workers are available and with enough lead time.					
Management from the host employer and staffing agencies are available to address day-to-day coordination issues related to safety.					

ESSENTIALS OF SAFETY

OSHA's Form 301

Injury and Illness Incident Report

Attention: This form contains information relating to employee health and must be used in a manner that protects the confidentiality of employees to the extent possible while the information is being used for occupational safety and health purposes.

U.S. Department of Labor
Occupational Safety and Health Administration

Form approved OMB no. 1218-0176

This *Injury and Illness Incident Report* is one of the first forms you must fill out when a recordable work-related injury or illness has occurred. Together with the *Log of Work-Related Injuries and Illnesses* and the accompanying *Summary*, these forms help the employer and OSHA develop a picture of the extent and severity of work-related incidents.

Within 7 calendar days after you receive information that a recordable work-related injury or illness has occurred, you must fill out this form or an equivalent. Some state workers' compensation, insurance, or other reports may be acceptable substitutes. To be considered an equivalent form, any substitute must contain all the information asked for on this form.

According to Public Law 91-596 and 29 CFR 1904, OSHA's recordkeeping rule, you must keep this form on file for 5 years following the year to which it pertains.

If you need additional copies of this form, you may photocopy and use as many as you need.

Completed by ______________________

Title ______________________

Phone (____) ____-______ Date ____/____/____

Information about the employee

1) Full name ______________________

2) Street ______________________

City ____________ State ______ ZIP ________

3) Date of birth ____/____/____

4) Date hired ____/____/____

5) ❒ Male
❒ Female

Information about the physician or other health care professional

6) Name of physician or other health care professional ______________________

7) If treatment was given away from the worksite, where was it given?

Facility ______________________

Street ______________________

City ____________ State ______ ZIP ________

8) Was employee treated in an emergency room?
❒ Yes
❒ No

9) Was employee hospitalized overnight as an in-patient?
❒ Yes
❒ No

Information about the case

10) Case number from the *Log* ____________ *(Transfer the case number from the Log after you record the case.)*

11) Date of injury or illness ____/____/____

12) Time employee began work ____________ AM / PM

13) Time of event ____________ AM / PM ❒ Check if time cannot be determined

14) ***What was the employee doing just before the incident occurred?*** Describe the activity, as well as the tools, equipment, or material the employee was using. Be specific. *Examples:* "climbing a ladder while carrying roofing materials"; "spraying chlorine from hand sprayer"; "daily computer key-entry."

15) ***What happened?*** Tell us how the injury occurred. *Examples:* "When ladder slipped on wet floor, worker fell 20 feet"; "Worker was sprayed with chlorine when gasket broke during replacement"; "Worker developed soreness in wrist over time."

16) ***What was the injury or illness?*** Tell us the part of the body that was affected and how it was affected; be more specific than "hurt," "pain," or "sore." *Examples:* "strained back"; "chemical burn, hand"; "carpal tunnel syndrome."

17) ***What object or substance directly harmed the employee?*** *Examples:* "concrete floor"; "chlorine"; "radial arm saw." *If this question does not apply to the incident, leave it blank.*

18) ***If the employee died, when did death occur?*** Date of death ____/____/____

Public reporting burden for this collection of information is estimated to average 22 minutes per response, including time for reviewing instructions, searching existing data sources, gathering and maintaining the data needed, and completing and reviewing the collection of information. Persons are not required to respond to the collection of information unless it displays a current valid OMB control number. If you have any comments about this estimate or any other aspects of this data collection, including suggestions for reducing this burden, contact: US Department of Labor, OSHA Office of Statistical Analysis, Room N-3644, 200 Constitution Avenue, NW, Washington, DC 20210. Do not send the completed forms to this office.

ESSENTIALS OF **SAFETY**

OSHA's Form 300 (Rev. 01/2004)

Log of Work-Related Injuries and Illnesses

U.S. Department of Labor • Occupational Safety and Health Administration

You must record information about every work-related death and about every work-related injury or illness that involves loss of consciousness, restricted work activity or job transfer, days away from work, or medical treatment beyond first aid. You must also record significant work-related injuries and illnesses that are diagnosed by a physician or licensed health-care professional. You must also record work-related injuries and illnesses that meet any of the specific recording criteria listed in 29 CFR 1904.8 through 1904.12. Feel free to use two lines for a single case if you need to. You must complete an Injury and Illness Incident Report (OSHA Form 301) or equivalent form for each injury or illness recorded on this form. If you're not sure whether a case is recordable, call your local OSHA office for help.

Attention: *This form contains information relating to employee health and must be used in a manner that protects the confidentiality of employees to the extent possible while the information is being used for occupational safety and health purposes.*

Company Name: ____________

Establishment Name: ____________

Address: ____________

City: ____________ State: ______ Zip Code: ______

Year 20___ ___

Page___ of___

Form approved OMB No. 1218-0176

See OMB disclosure statement on reverse.

IDENTIFY THE PERSON			DESCRIBE THE CASE			CLASSIFY THE CASE — CHECK ONLY ONE box for each case based on the most serious outcome for that case:				Enter the number of days the injured or ill worker was:		Check the "Injury" column or choose one type of illness: (M)					
						Death	Days away from work	Remained at work									
(A) Case no.	(B) Employee's name	(C) Job title (e.g., Welder)	(D) Date of injury or onset of illness	(E) Where the event occurred (e.g., Loading dock north end)	(F) Describe injury or illness, parts of body affected, and object/substance that directly injured or made person ill (e.g., Second degree burns on right forearm from acetylene torch)	(G)	(H)	Job transfer or restriction (I)	Other record-able cases (J)	Away from work (K)	On job transfer or restriction (L)	Injury (1)	Skin disorder (2)	Respiratory condition (3)	Poisoning (4)	Hearing loss (5)	All other illnesses (6)
			month / day			☐	☐	☐	☐	___ days	___ days	☐	☐	☐	☐	☐	☐
			month / day			☐	☐	☐	☐	___ days	___ days	☐	☐	☐	☐	☐	☐
			month / day			☐	☐	☐	☐	___ days	___ days	☐	☐	☐	☐	☐	☐
			month / day			☐	☐	☐	☐	___ days	___ days	☐	☐	☐	☐	☐	☐
			month / day			☐	☐	☐	☐	___ days	___ days	☐	☐	☐	☐	☐	☐
			month / day			☐	☐	☐	☐	___ days	___ days	☐	☐	☐	☐	☐	☐
			month / day			☐	☐	☐	☐	___ days	___ days	☐	☐	☐	☐	☐	☐
			month / day			☐	☐	☐	☐	___ days	___ days	☐	☐	☐	☐	☐	☐
			month / day			☐	☐	☐	☐	___ days	___ days	☐	☐	☐	☐	☐	☐
			month / day			☐	☐	☐	☐	___ days	___ days	☐	☐	☐	☐	☐	☐
			month / day			☐	☐	☐	☐	___ days	___ days	☐	☐	☐	☐	☐	☐
			month / day			☐	☐	☐	☐	___ days	___ days	☐	☐	☐	☐	☐	☐
			month / day			☐	☐	☐	☐	___ days	___ days	☐	☐	☐	☐	☐	☐
			month / day			☐	☐	☐	☐	___ days	___ days	☐	☐	☐	☐	☐	☐
			month / day			☐	☐	☐	☐	___ days	___ days	☐	☐	☐	☐	☐	☐
			month / day			☐	☐	☐	☐	___ days	___ days	☐	☐	☐	☐	☐	☐
			month / day			☐	☐	☐	☐	___ days	___ days	☐	☐	☐	☐	☐	☐
			month / day			☐	☐	☐	☐	___ days	___ days	☐	☐	☐	☐	☐	☐
			month / day			☐	☐	☐	☐	___ days	___ days	☐	☐	☐	☐	☐	☐
			month / day			☐	☐	☐	☐	___ days	___ days	☐	☐	☐	☐	☐	☐
					Page totals ⇨	___ (G)	___ (H)	___ (I)	___ (J)	___ (K)	___ (L)	___ (1)	___ (2)	___ (3)	___ (4)	___ (5)	___ (6)

Do Not Post This Form! Be sure to transfer these totals to the Summary page (Form 300A) before you post it.

Retain and update until ______ / ______ / 20___ ___ (5 years after the end of the current year. See §1904.33)

Do not send completed forms to OSHA unless requested – See §1904.41. See Reverse Side For 300 Log Instructions. All entries need to have a corresponding OSHA 301 Injury and Illness Incident Report or an equivalent form completed. See §1904.29(b)(2) A free copy of the Annual Summary, First Report Of Injury – OSHA Form No. 300A and 301, respectively, may be downloaded from www.mancomm.com.

OSHA's Form 300A (Rev. 01/2004)

Summary of Work-Related Injuries and Illnesses

Year 20__ __

U.S. Department of Labor
Occupational Safety and Health Administration

Form approved OMB no. 1218-0176

All establishments covered by Part 1904 must complete this Summary page, even if no work-related injuries or illnesses occurred during the year. Remember to review the Log to verify that the entries are complete and accurate before completing this summary.

Using the Log, count the individual entries you made for each category. Then write the totals below, making sure you've added the entries from every page of the Log. If you had no cases, write "0."

Employees, former employees, and their representatives have the right to review the OSHA Form 300 in its entirety. They also have limited access to the OSHA Form 301 or its equivalent. See 29 CFR Part 1904.35, in OSHA's recordkeeping rule, for further details on the access provisions for these forms.

Number of Cases

Total number of deaths	Total number of cases with days away from work	Total number of cases with job transfer or restriction	Total number of other recordable cases
______	______	______	______
(G)	(H)	(I)	(J)

Number of Days

Total number of days away from work	Total number of days of job transfer or restriction
______	______
(K)	(L)

Injury and Illness Types

Total number of . . .
(M)

(1) Injuries	______	(4) Poisonings	______
(2) Skin disorders	______	(5) Hearing Losses	______
(3) Respiratory conditions	______	(6) All other illnesses	______

Post this Summary page from February 1 to April 30 of the year following the year covered by the form.

Public reporting burden for this collection of information is estimated to average 50 minutes per response, including time to review the instructions, search and gather the data needed, and complete and review the collection of information. Persons are not required to respond to the collection of information unless it displays a currently valid OMB control number. If you have any comments about these estimates or any other aspects of this data collection, contact: US Department of Labor, OSHA Office of Statistical Analysis, Room N-3644, 200 Constitution Avenue, NW, Washington, DC 20210. Do not send the completed forms to this office.

Establishment information

Your establishment name ______________________

Street ______________________

City ____________ State ______ ZIP ______

Industry description (*e.g., Manufacture of motor truck trailers*)

____

Standard Industrial Classification (SIC), if known (*e.g., 3715*)

__ __ __ __

OR

North American Industrial Classification (NAICS), if known (*e.g., 336212*)

__ __ __ __ __ __

Employment information

(If you don't have these figures, see the Worksheet on the back of this page to estimate.)

Annual average number of employees ______

Total hours worked by all employees last year ______

Sign here

Knowingly falsifying this document may result in a fine.

I certify that I have examined this document and that to the best of my knowledge the entries are true, accurate, and complete.

____
Company executive   Title

( ) - / /
Phone   Date

HEPATITIS B VACCINE DECLINATION

MANCOMM®

I understand that due to my occupational exposure to blood or other potentially infectious materials, I may be at risk of acquiring hepatitis B virus (HBV) infection.

I have been given the opportunity to be vaccinated with hepatitis B vaccine, at no charge to myself. However, I decline hepatitis B vaccination at this time. I understand that by declining this vaccine, I continue to be at risk of acquiring hepatitis B, a serious disease.

If in the future I continue to have occupational exposure to blood or other potentially infectious materials and I want to be vaccinated with hepatitis B vaccine, I can receive the vaccination series at no charge to me.

I certify by signing this document that I have read and understood the above information.

______________________	______________________
Employee's Signature	Print Name
______________________	______________________
Date	Company Name

Example of Authorized Labeling

BIOHAZARD

[Or, in the case of regulated waste, the legend:]

BIOHAZARD WASTE

[as described in 29 CFR 1910.1030]

These labels must be fluorescent orange or orange-red, or predominantly so, with lettering and symbols in a contrasting color.

SHARPS INJURY LOG[1]

MANCOMM®

For Period Ending: ____ / ____ / ________

Company Name: ______________________________

Date Entered:	Date Incident Occurred & Time Incident Occured:	Type and Brand of Device Involved:	Department or Work Area Where Exposure Incident Occurred:	How Incident Occurred:
___/___/___ Month Day Year	___/___/___ Month Day Year ___:___ Hour Minute AM PM			
___/___/___ Month Day Year	___/___/___ Month Day Year ___:___ Hour Minute AM PM			
___/___/___ Month Day Year	___/___/___ Month Day Year ___:___ Hour Minute AM PM			
___/___/___ Month Day Year	___/___/___ Month Day Year ___:___ Hour Minute AM PM			
___/___/___ Month Day Year	___/___/___ Month Day Year ___:___ Hour Minute AM PM			
___/___/___ Month Day Year	___/___/___ Month Day Year ___:___ Hour Minute AM PM			
___/___/___ Month Day Year	___/___/___ Month Day Year ___:___ Hour Minute AM PM			
___/___/___ Month Day Year	___/___/___ Month Day Year ___:___ Hour Minute AM PM			
___/___/___ Month Day Year	___/___/___ Month Day Year ___:___ Hour Minute AM PM			
___/___/___ Month Day Year	___/___/___ Month Day Year ___:___ Hour Minute AM PM			
___/___/___ Month Day Year	___/___/___ Month Day Year ___:___ Hour Minute AM PM			

- **Retain until** ______ / ______ / ______ (5 years after the end of the current year - see §1904.44)
- You are required to maintain this log if the requirement to maintain a 300 log applies to you. See Part 1904.

[1] **Referred to in §1910.1030(h)(5)**

Module 1 - Introduction to OSHA

1. OSHA began because, until ________, there were no national laws regulating safety and health hazards.

   (a) 1910

   (b) 1940

   (c) 1970

   (d) 1990

2. Nearly every American employee is covered by OSHA or an OSHA-approved state agency.

   ____ True ____ False

3. Which of the following is **not** a right every employee has under OSHA?

   (a) the right to a safe and healthful workplace

   (b) the right to complain or request hazard correct from your employer

   (c) the right to file a complaint with OSHA

   (d) the right to refuse to work because you just don't feel like it

4. Which of the following personal protective equipment must an employer pay for?

   (a) a hard hat that the employee intentionally smashed with a hammer

   (b) steel-toe boots that must be left in the employee's locker when he or she is not working

   (c) steel-toe boots that the employer allows to be worn off the jobsite

   (d) sunscreen

5. Section 5(a)(1) of the OSH Act, which states "Each employer shall furnish to each of his employees employment and a place of employment which are free from recognized hazards that are causing or are likely to cause death or serious physical harm to his employees," is called ____________________.

   (a) the General Duty Clause

   (b) the Safe and Healthful Workplace Clause

   (c) the Employer's Duty Clause

   (d) none of the above

Module 2 - RegLogic™

1. What does "CFR" stand for?
   (a) Coded Federal Register
   (b) Code of Federal Regulations
   (c) Civil Federated Regulations
   (d) Code of First Regulations

2. Which part of the CFR contains the regulations for recording and reporting occupational injuries and illnesses?
   (a) 1926
   (b) 1910
   (c) 1904
   (d) 1903

3. Part 1926 contains the occupational safety and health standards for which industry?
   (a) general industry
   (b) maritime
   (c) construction
   (d) agriculture

4. Which CFR title contains the regulations applicable to labor?
   (a) Title 15
   (b) Title 29
   (c) Title 50
   (d) Title 1910

5. OSHA follows a distinct pattern in the way it presents information; therefore, training requirements are always found in the same paragraph throughout the different standards.

   ____ True ____ False

Module 3 - General Safety and Health Provisions

1. Employers may permit only those employees qualified by __________________ to operate equipment or machinery.

   (a) training

   (b) experience

   (c) either (a) or (b)

   (d) none of the above

2. Waste containers for flammable or hazardous wastes must be covered.

   ____True ____ False

3. A___________________ is one who, by possession of a recognized degree, certificate, or professional standing, or who by extensive knowledge, training, and experience, has successfully demonstrated his or her ability to solve or resolve problems relating to the subject matter, the work, or the project.

   (a) "qualified person"

   (b) "competent person"

   (c) manager

   (d) supervisor

4. Which of the following is **not** true about drinking water?

   (a) An adequate supply of potable drinking water must be provided in all places of employment.

   (b) Dipping and/or using common drinking cups is permitted.

   (c) Portable containers for drinking water must be capable of being tightly closed and be equipped with a tap.

   (d) Drinking water containers must be marked as such.

5. If a construction jobsite has 40 workers, what is the minimum number of toilet facilities that must be provided?

   (a) two toilet seats and two urinals

   (b) one toilet seat

   (c) one urinal

   (d) one toilet seat and one urinal

Module 4 - Fall Protection

1. The general rule is that fall protection is required whenever a worker could fall ____________ or more.

   (a) 5 feet

   (b) 6 feet

   (c) 10 feet

   (d) 12 feet

2. How far from the walking/working surface must the top rail of a guardrail system used for fall protection be?

   (a) 24 – 28 inches

   (b) 32 – 36 inches

   (c) 39 – 45 inches

   (d) 48 – 52 inches

3. Safety nets must be installed as close as practicable under the walking/working surface, but never more than 20 feet below the working level.

   ____ True ____ False

4. A personal fall arrest system must limit the maximum deceleration distance an employee travels to _______.

   (a) 3.5 feet

   (b) 5 feet

   (c) 6 feet

   (d) 7.5 feet

5. __________________ may only be four times as tall as their shortest base dimension.

   (a) Ladders

   (b) Aerial lifts

   (c) Positioning device systems

   (d) Supported scaffolds

Module 5 - Electrocution Hazards

1. A/an ________________ results from the body becoming part of an electrical circuit, with current entering the body at one point and leaving at another.

   (a) arc flash or blast

   (b) fire

   (c) shock

   (d) explosion

2. The covering on an overhead power line is primarily for protection from shock.

   ____ True ____ False

3. What is the minimum required clearance distance from a power line with a voltage of 100 kV?

   (a) 10 feet

   (b) 15 feet

   (c) 20 feet

   (d) 25 feet

4. Receptacle ground fault circuit interrupters should be tested:

   (a) after installation

   (b) prior to each and every use

   (c) once a month

   (d) both (a) and (c)

5. Which of the following is **not** good practice when using power tools in order to avoid electrocution hazards?

   (a) Always carry power tools by their cords.

   (b) Keep cords away from heat, oil, and sharp edges.

   (c) Do not use tools in wet or damp environments.

   (d) Use double-insulated tools.

Module 6 - Caught-in or -Between Hazards

1. Which of the following is **not** an example of a caught-in or -between hazard?
   (a) being caught in a cave-in during trenching operations
   (b) being struck on the head by a falling tool
   (c) being crushed between a wall and an I-beam sliding off of a truck
   (d) being pulled into machinery

2. If machines are not deenergized and locked out when they are being repaired, they may cycle or otherwise start up and catch a worker's body part or clothing and cause injury or death.

   ____ True ___ False

3. Which of the following must be guarded if it is exposed to contact by workers?
   (a) pulley
   (b) spindle or drum
   (c) fly wheel
   (d) all of the above

4. In order to ensure that it is sufficiently supported, secured, or otherwise made safe, you should always raise the blades of bulldozers, scrapers, and similar equipment when the equipment is not in use.

   ____True ____ False

5. In order to protect yourself from caught-in or -between hazards on an excavation site, never work in an unprotected trench that is___________ deep or more.
   (a) 5 feet
   (b) 8 feet
   (c) 10 feet
   (d) 12 feet

Module 7 - Struck-by Hazards

1. A piece of a drill bit that separates from the tool and strikes the tool operator would be an example of a ________________________ hazard.

   (a) struck by a rolling object

   (b) struck by a flying object

   (c) struck by a falling object

   (d) struck by a swinging object

2. Which of the following should you do in order to protect yourself from struck-by hazards when working near heavy equipment that is in operation?

   (a) Never work under a suspended load.

   (b) Confirm and receive acknowledgement from the heavy equipment operator that you are visible.

   (c) Be aware of the swing radius of cranes and backhoes, and do not enter that zone.

   (d) Do all of the above.

3. Compressed air must be reduced to ________ psi or less of air pressure before it may be used for cleaning purposes.

   (a) 50

   (b) 45

   (c) 30

   (d) 15

4. In order to avoid struck-by hazards, which of the following should a worker be prohibited from using?

   (a) a file where the manufacturer's name has been worn away

   (b) a screwdriver with a splintered handle

   (c) a hammer with a mushroomed head

   (d) both (b) and (c)

5. When performing overhead work, you should ensure that all materials stored in buildings under construction are placed farther than ________ away from hoist way or floor openings and more than __________ from exterior walls.

   (a) 6 feet / 10 feet

   (b) 10 feet / 6 feet

   (c) 4 feet / 8 feet

   (d) 8 feet / 4 feet

Module 8 - Personal Protective Equipment (PPE)

1. Which of the following is **not** true about personal protective equipment (PPE)?
   (a) Employees must pay for their own PPE.
   (b) PPE must be provided, used, and maintained in a sanitary and reliable condition.
   (c) PPE must be of a safe design and construction for the work to be performed.
   (d) PPE must not be used if it is defective or damaged.

2. Replacement PPE must be paid for by the employer, except when an employee loses or intentionally damages their PPE.
   ____True ____ False

3. What is the "last line of defense" against work hazards?
   (a) engineering controls
   (b) safe work practices
   (c) PPE
   (d) none of the above

4. If you are required to use PPE, you must be trained to know:
   (a) when and which PPE is necessary, and the limitations of that PPE
   (b) how to properly put on, take off, adjust, and wear the PPE
   (c) the proper care, maintenance, useful life, and disposal of the PPE
   (d) all of the above

5. Which class of head protection protects against falling objects and high-voltage shocks/burns?
   (a) Class G
   (b) Class E
   (c) Class C
   (d) Class A

Module 9 - Health Hazards in Construction

1. Permanent lung damage caused by breathing dust containing extremely fine particles of crystalline silica is called ____________________.

   (a) heat stress

   (b) asbestosis

   (c) silicosis

   (d) lead poisoning

2. Using a vacuum with a HEPA filter will help to prevent exposure to harmful ____________ dust.

   (a) silica

   (b) asbestos

   (c) cadmium

   (d) lead

   (e) all of the above

3. Who is responsible for determining if asbestos is present in the work environment and, if so, if the amount present exceeds the permissible exposure limit (PEL)?

   (a) the employee

   (b) the employer

   (c) OSHA

   (d) both the employee and the employer

4. Regulated areas must be established when employees are reasonably expected to be exposed to cadmium beyond the PEL.

   ____True ____ False

5. The PEL for lead is 50 $\mu g/m^3$ averaged over an eight-hour period. The __________________ , or the level at which an employer must begin the specific compliance activities outlined in the OSHA standard, for lead is an airborne concentration of 30 $\mu g/m^3$ averaged over an eight-hour period.

   (a) action level

   (b) regulatory level

   (c) permissible exposure level

   (d) none of the above

Module 10 - Hazard Communication, Part 1

1. HazCom is also called "the ___________ to Know."
   (a) Right
   (b) Duty
   (c) Privilege
   (d) Chance
2. If a chemical is flammable or explosive, it poses a _____________.
   (a) simple asphyxiant hazard
   (b) health hazard
   (c) physical hazard
   (d) hazard not otherwise classified
3. A _____________________ is a statement assigned to a hazard class and category that describes the nature of the chemical's hazards including, where appropriate, the degree of hazard.
   (a) precautionary statement
   (b) hazard statement
   (c) signal word
   (d) pictogram
4. You would find a description of the most important symptoms or effects of exposure, both acute and delayed, in Section 8, Exposure Controls/Personal Protection, of a chemical's safety data sheet.
   ____True ____ False
5. Which of the following pictogram descriptions is **not** correctly paired with the hazard it represents?
   (a) Flame over Circle – organic peroxide
   (b) Skull & Crossbones – acute toxicity
   (c) Gas Cylinder – gas under pressure
   (d) Exclamation Mark – narcotic effects

Module 11 - Cranes and Derricks

1. A/an ____________________must begin a visual inspection for apparent deficiencies prior to each shift that a crane is used; this inspection must be completed before or during that shift.

   (a) supervisor

   (b) employee

   (c) competent person

   (d) qualified person

2. A crane operator is not allowed to leave the crane's controls unattended while a load is suspended unless:

   (a) the operator remains adjacent to the controls and is not engaged in other duties

   (b) the load is to be suspended longer than normal lifting operations

   (c) a competent person determines it is safe to do and implements the necessary safety measures

   (d) barricades or caution lines are set up to prevent employees from entering the fall zone

   (e) all of the above

3. Using crane equipment to hoist personnel is prohibited except for when the employer can demonstrate:

   (a) conventional means of reaching the work area is more dangerous

   (b) the project's structural design or worksite conditions prohibit the use of conventional means of reaching the work area

   (c) either (a) or (b)

   (d) neither (a) nor (b)

4. When using signals, all directions given to the operator by the signal person must be given from the ________________ direction perspective.

   (a) operator's

   (b) signal person's

   (c) competent person's

   (d) supervisor's

5. The crane operator must be informed when an employee enters the swing radius of the crane and is not permitted to rotate the superstructure of the crane until informed that the employee is in a safe position.

   ___True ___ False

Module 12 - Excavations

1. The excavation, the area around it, and the protective system(s) in use require inspection by a competent person only once a day.

   ____ True ____ False

2. Which of the following is **not** a common cause of cave-ins?

   (a) lack of protection system

   (b) spoils and equipment too far from the excavation

   (c) water accumulation

   (d) vibrations from large equipment

   (e) undermining of adjacent buildings, walls, or other structures

3. A protective system is **not** required for an excavation less than __________ deep if a competent person has decided that there is no danger of a cave-in.

   (a) 8 feet

   (b) 6 feet

   (c) 5 feet

   (d) 4 feet

4. Excavations more than four feet deep must be tested for ____________ before an employee enters the excavation.

   (a) oxygen deficiency

   (b) high concentrations of combustible gas

   (c) high levels of hazardous substances

   (d) all of the above

   (e) none of the above

5. Spoils must be placed more than __________ from the edge of the excavation, measured from the nearest part of the spoils to the excavation's edge.

   (a) 1 foot

   (b) 2 feet

   (c) 5 feet

   (d) 10 feet

Module 13 - Materials Handling and Storage

1. Do not stack lumber more than ___________ high when stacking manually.

   (a) 10 feet

   (b) 12 feet

   (c) 16 feet

   (d) 20 feet

2. Materials stored inside buildings under construction must be placed at least six feet from ________________________.

   (a) hoistways and floor openings

   (b) windows

   (c) exterior walls that do not extend above the top of the material stored

   (d) other materials

3. Stacks of ________________ may not be more than seven feet in height and must taper back two inches for every foot of height above the four-foot level.

   (a) lumber

   (b) bricks

   (c) bags or bundles

   (d) masonry blocks

4. Earthmoving equipment must have appropriate seat belts and rollover protective structures.

   ____ True ____ False

5. Wire rope shall not be used in rigging equipment if the total number of visible broken wires exceeds _____________ of the total number of wires.

   (a) 25%

   (b) 20%

   (c) 10%

   (d) 5%

Module 14 - Scaffolds

1. Which of the following types of scaffolds may **not** be used?

   (a) shore scaffolds (supported scaffolds that are held against a building or structure with props)

   (b) suspension scaffolds (platforms suspended by ropes or some other non-rigid means from an overhead structure)

   (c) lean-to scaffolds (supported scaffolds that are kept erect by leaning against a building or structure)

   (d) all of the above

   (e) both (a) and (c)

2. Each employee on a scaffold more than ___________ above a lower level must be protected from falling by a personal fall arrest system and/or guardrail system (or, for crawling boards, a grabline or equivalent handhold).

   (a) 6 feet

   (b) 10 feet

   (c) 12 feet

   (d) 20 feet

3. When working on a scaffold, you must maintain a minimum clear distance of ___________ from an overhead power line with a voltage of 60 kV.

   (a) 20 feet

   (b) 16 feet

   (c) 14 feet

   (d) 10 feet

4. Employees must wear hard hats when on a scaffold.

   ___ True ___ False

5. Which of the following is a common cause of falls from a scaffold?

   (a) lack of a fall protection system

   (b) slipping or tripping

   (c) improper means of climbing on or off the scaffold

   (d) improper construction or failure of the scaffold

   (e) all of the above

Module 15 - Stairways and Ladders

1. Stairways or ladders are rarely the source of injuries serious enough to require time off the job.

   ____ True ____ False

2. There must be a stairway or ladder at every point of access where there is a break in elevation of _______________ or more.

   (a) 12 inches

   (b) 15 inches

   (c) 19 inches

   (d) 24 inches

3. Every stairway ________________________ must have at least one handrail and a stairrail along each unprotected edge.

   (a) with four or more risers or that rises more than 30 inches, whichever is less

   (b) with three or more risers or that rises more than 24 inches, whichever is less

   (c) with five or more risers

   (d) that rises more than 24 inches

4. Non-self-supporting ladders that lean against a wall or other support should be positioned at an angle where the horizontal distance from the top support to the foot of the ladder is________ the working length of the ladder.

   (a) 3/4

   (b) 2/3

   (c) 1/3

   (d) 1/4

5. How far must the side rails of a portable ladder used for access to an upper landing surface extend above the upper landing surface?

   (a) 4 feet

   (b) 3 feet

   (c) 2 feet

   (d) 1 foot

Module 16 - Tools

1. Which of the following poses the greatest hazard when using hand tools?
   (a) misuse
   (b) improper maintenance
   (c) manufacturing flaws
   (d) both (a) and (b)
   (e) both (a) and (c)

2. Dull tools can cause more hazards than sharp ones.
   ____ True ____ False

3. Pneumatic tools that shoot nails, rivets, staples, or similar fasteners and operate at pressures of more than ____________ must be equipped with a special device to keep fasteners from being ejected unless the muzzle is pressed against the work surface.
   (a) 25 p.s.i.
   (b) 50 p.s.i.
   (c) 100 p.s.i.
   (d) 125 p.s.i.

4. Never load a powder-actuated tool unless it is to be used ____________.
   (a) immediately
   (b) within 5 minutes
   (c) within 15 minutes
   (d) within a reasonable amount of time

5. Abrasive wheel tools must be equipped with guards that cover the ________________.
   (a) spindle end
   (b) nut
   (c) flange projections
   (d) all of the above

Module 17 - Concrete and Masonry Construction

1. Construction loads may not be placed on a concrete structure or portion thereof unless the employer determines, based on information received from a/an ___________________________, that the structure or portion thereof is capable of supporting the intended load.

   (a) employee

   (b) competent person

   (c) supervisor

   (d) person qualified in structural design

2. Employees (except those essential to the post-tensioning operation) are not permitted to be in front of the jack during tensioning operations.

   ____ True ____ False

3. Employees are not permitted to apply a cement, sand, and water mixture through a pneumatic hose unless:

   (a) they are wearing protective head and face equipment

   (b) they are wearing protective clothing

   (c) a supervisor gives permission

   (d) OSHA gives permission

4. Which of the following is **not** true about jacking equipment used in lift-slab operations?

   (a) The jacking equipment must be marked with the manufacturer's rated capacity.

   (b) The jacking equipment must be capable of supporting at least two-and-a-half times the load being lifted during the jacking operation.

   (c) The jacking equipment must not be overloaded.

   (d) All of the above are true.

5. The limited access zone that must be established around a masonry wall that is under construction must be equal to the height of the wall to be constructed plus _________.

   (a) 2 feet

   (b) 4 feet

   (c) 6 feet

   (d) 8 feet

Module 18 - Confined Space Entry

1. A ____________________ is large enough and so configured that an employee can enter and perform assigned work, has limited or restricted means for entry and exit, is not designed for continuous employee occupancy, and contains a hazard.

   (a) confined space

   (b) permit-required confined space

   (c) dangerous space

   (d) regulated space

2. Whose responsibility is it to maintain a continuous, accurate account of all authorized entrants in a permit-required confined space?

   (a) the entry supervisor

   (b) rescue personnel

   (c) the attendant

   (d) the authorized entrants

3. Canceled entry permits for permit-required confined spaces must be retained for how long?

   (a) at least 1 month

   (b) at least 1 year

   (c) at least 10 years

   (d) Canceled permits do not have to be retained.

4. Unless the retrieval equipment would increase the overall risk of entry, a retrieval system must be used whenever an authorized entrant enters a permit space. A mechanical device must be available to retrieve personnel from vertical-type permit spaces that are more than 5 feet deep.

   ___ True ___ False

5. Which of the following would **not** be found on an entry permit for a permit-required confined space?

   (a) the qualifications of all authorized entrants

   (b) the names of all authorized entrants, personnel serving as attendants, and the entry supervisor

   (c) initial and periodic monitoring results

   (d) the measures used to isolate the permit space, as well as eliminate and control hazards

Module 19 - Fire Protection and Prevention

1. A fire extinguisher rated not less than 2A or a 55-gallon open drum of water with two fire pails must be provided for each ______________ of protected building area.

   (a) 1,000 sq. feet

   (b) 2,000 sq. feet

   (c) 3,000 sq. feet

   (d) 5,000 sq. feet

2. Which of the following is true about the fire extinguishers required on construction sites?

   (a) Two or more fire extinguishers, rated 2A, must be provided on each floor.

   (b) At construction sites involving multistory buildings, at least one fire extinguisher must be located adjacent to the stairway.

   (c) If more than three gallons of flammable or combustible liquid is present, a fire extinguisher rated not less than 2A must be provided within 100 feet.

   (d) Carbon tetrachloride fire extinguishers may be used to meet the OSHA requirements.

3. With reference to fire extinguishers, what does the acronym "PASS" stand for?

   (a) Pull, Aim, Squeeze, Sweep

   (b) Point, Aim, Sweep, Squeeze

   (c) Pass, Around, Safety, Side

   (d) Pull, Alarm, Safety, Sweep

4. No more than ___________ of Category 1, 2, or 3 flammable liquids or ___________ of Category 4 flammable liquids may be stored in any one storage cabinet.

   (a) 120 gallons; 90 gallons

   (b) 60 gallons; 120 gallons

   (c) 60 gallons; 90 gallons

   (d) 90 gallons; 60 gallons

5. No more than 500 gallons of flammable liquids may be stored in containers in any one outdoor pile.

   ____ True ____ False

Module 20 - Steel Erection

1. Before steel erection may begin, the controlling contractor must provide the steel erector with written notification of what?

   (a) that the concrete in the footings, piers, and walls has sufficient strength to support the loads imposed during steel erection

   (b) that the mortar in the masonry piers has sufficient strength to support the load imposed during steel erection

   (c) that any repairs or modification to anchor bolts were properly conducted

   (d) all of the above

   (e) none of the above

2. What is the maximum number of members that may be lifted when performing a multiple lift or "Christmas treeing?"

   (a) 2 members

   (b) 5 members

   (c) 7 members

   (d) 10 members

3. Shear connectors on composite floors, roofs, or bridge decks may be installed from within a controlled decking zone.

   ____ True ____ False

4. What is the minimum number of anchor bolts/rods that must be used to anchor columns?

   (a) 8 bolts/rods

   (b) 6 bolts/rods

   (c) 4 bolts/rods

   (d) 2 bolts/rods

5. Each employee engaged in a steel erection activity who is on a walking/working surface with an unprotected side or edge more than __________ above a lower level must be protected by a guardrail, safety net, personal fall arrest, positioning device, or fall restraint system.

   (a) 6 feet

   (b) 10 feet

   (c) 15 feet

   (d) 20 feet

Module 21 - Welding and Cutting

1. Which of the following is **not** true about transporting compressed gas cylinders?
   (a) Valve protections caps must be in place and secured.
   (b) Cylinders should always be stored on their side.
   (c) Cylinders should be moved by tilting and rolling.
   (d) When being transported by vehicle, cylinders must be secured at all times.

2. A cylinder must be "cracked," or opened and closed quickly while standing to the side to clear the valve of dust, before connecting it to a regulator.
   ____ True ____ False

3. Which of the following is true about arc welding operations?
   (a) Electrode holders that are not in use must be placed so that they cannot make electrical contact with persons, conducting objects, fuel, or compressed gas tanks.
   (b) Electrode holds should not be dipped in water.
   (c) Cables with splices within 10 feet of the holder must not be used.
   (d) Only qualified personnel are permitted to repair arc welding equipment.
   (e) All of the above are true.

4. In an enclosed space when welding, cutting, or heating on a surface coated with toxic preservatives, how must employees be protected from hazardous fumes?
   (a) All toxic coatings on the surface must be stripped for a distance of at least four inches from the area of heat application.
   (b) Employees must wear air line respirators.
   (c) either (a) or (b)
   (d) both (a) and (b)
   (e) neither (a) nor (b)

5. All welding and cutting operations carried on in confined spaces must be adequately ventilated to prevent the accumulation of toxic materials or possible ______________.
   (a) fire hazards
   (b) oxygen deficiency
   (c) arc blasts
   (d) none of the above

Module 22 - Motor Vehicles

1. Equipment must have appropriate lights or reflectors (or barricades with lights or reflectors) whenever it is:

   (a) left unattended at night

   (b) adjacent to a highway in normal use

   (c) adjacent to construction areas when work is in progress

   (d) all of the above

   (e) none of the above

2. Vehicles are required to have operable brake lights only when the vehicle is being used at night or during low-light conditions.

   ____ True ____ False

3. Motor vehicle equipment that has an obstructed view to the rear must have what?

   (a) a flashing light on its hood

   (b) oversized side mirrors and an oversized rearview mirror

   (c) a reverse signal alarm that is audible above the surrounding noise level

   (d) all of the above

   (e) none of the above

4. OSHA requirements concerning which of the following do **not** specifically incorporate the requirements of Part VI of the Federal Highway Administration's Manual on Uniform Traffic Control Devices (MUTCD)?

   (a) flaggers and the warning garments they wear

   (b) site-specific requirements

   (c) barricades

   (d) traffic control signs/devices and signaling

5. The retroreflective material on the high-visibility safety apparel that flaggers are required to wear must be visible at a minimum distance of ______________.

   (a) 350 feet

   (b) 500 feet

   (c) 1,000 feet

   (d) 2,500 feet

Module 23 - Powered Industrial Trucks

1. Modifications to powered industrial trucks that affect capacity and safe operation may not be performed without prior written approval from ________________.

   (a) the manufacturer

   (b) OSHA

   (c) a competent and qualified mechanic

   (d) a supervisor

2. Which of the following is **not** true about changing and/or charging powered industrial truck batteries?

   (a) Facilities must be provided for flushing and neutralizing spilled electrolyte.

   (b) Smoking is permitted in the charging area.

   (c) Eyewash facilities must be readily available in the designated battery changing area.

   (d) Adequate ventilation for the dispersal of vapors from gassing batteries must be provided.

3. Refresher training or an evaluation of a powered industrial truck operator's previous training is required at least every three years.

   ____ True ____ False

4. At least how old must forklift operators be under federal labor regulations?

   (a) There is no age requirement.

   (b) 21

   (c) 18

   (d) 16

5. What must happen when a powered industrial truck is left unattended?

   (a) The load-engaging means must be fully lowered.

   (b) The controls must be neutralized and the power shut off.

   (c) The brakes must be set; if the truck is parked on an incline, the wheels must also be blocked.

   (d) All of the above must occur.

Module 24 - DOT Placarding

1. DOT markings must be maintained on containers until:
   (a) the container is cleaned of residue
   (b) the container is purged of vapors
   (c) both (a) and (b)
   (d) neither (a) nor (b)

2. EPA requires containers to be _______________ to remove all hazardous characteristics from the container.
   (a) rinsed
   (b) double-rinsed
   (c) triple-rinsed
   (d) none of the above

3. Any employer who receives a shipment required to be placarded in accordance with the hazardous materials regulations must retain those placards until:
   (a) the hazardous materials are sufficiently removed to prevent any potential hazards
   (b) OSHA gives written approval to remove then
   (c) until EPA gives written approval to remove them
   (d) 60 calendar days have passed

4. Markings, placards, and labels must be readily visible.
   ____ True ____ False

5. For _______________________, the DOT label can be removed if another label is affixed in accordance with OSHA's Hazard Communication Standard.
   (a) bulk packages that will be reshipped
   (b) non-bulk packages that will not be reshipped
   (c) non-bulk packages that will be reshipped
   (d) none of the above

Module 25 - Electrical

1. Conductors and equipment are acceptable only if approved by ______________.
   (a) OSHA
   (b) a qualified national testing facility
   (c) a licensed electrician
   (d) the employer
2. Which of the following must be grounded?
   (a) supports and enclosures, metal cable trays, metal raceways, and mental enclosures for conductors
   (b) exposed, non-current-carrying metal parts of fixed equipment that may become energized
   (c) exposed, non-current-carrying metal parts of cord- and plug-connected equipment that may become energized
   (d) all of the above
3. All pull boxes, junction boxes, and fittings must be provided with covers approved for that purpose; if metal covers are used, they must be grounded.
   ____ True ____ False
4. Flexible cords and cables used with temporary and portable lights must be designed for _____________ usage.
   (a) light
   (b) medium
   (c) hard
   (d) extra-hard
   (e) either (c) or (d)
5. Which of the following is acceptable?
   (a) a flexible cable used to connect a portable appliance
   (b) a flexible cord run through a window to power a saw
   (c) a flexible cord installed in a raceway
   (d) a flexible cord running from the second floor down to the first through a stairway opening

Module 26 - Lockout/Tagout

1. Under OSHA's requirements for construction, controls that are to be deactivated during the course of work on energized or deenergized equipment or circuits must be ________________ .

   (a) discussed

   (b) tagged

   (c) selected by a licensed electrician

   (d) none of the above

2. Under the general industry requirements, control circuit devices (e.g., push buttons, selector switches, or interlocks) should be the sole means for deenergizing circuits or equipment.

   ___ True ___ False

3. Under the general industry requirements, when may a tag be used without a lock?

   (a) if a lock cannot be applied

   (b) if the employer can demonstrate that tagging procedures will provide a level of safety equivalent to that obtained by use of a lock

   (c) either (a) or (b)

   (d) neither (a) nor (b)

4. Before re-energizing equipment or circuits after lockout/tagout, what should happen?

   (a) A qualified person should conduct tests and visual inspections to verify that all tools, electrical jumpers, shorts, grounds, and other such devices have been removed so that the circuits and equipment can be safely energized.

   (b) Employees exposed to potential start-up hazards should be warned to stay clear of circuits and equipment.

   (c) Each lock and tag should be removed only by the employee who applied it.

   (d) There should be a visual determination that all employees are clear of the circuits and equipment.

   (e) All of the above should happen before re-energizing the equipment or circuits.

5. Under OSHA's general industry requirements, which type of employee should receive training on recognition of hazardous energy sources, the type and magnitude of the energy involved, and the methods necessary for energy isolation and control?

   (a) authorized employees

   (b) affected employees

   (c) other employees

   (d) none of the above, as lockout/tagout training requirements for general industry do not exist

Module 27 - Signs/Tags

1. Which type of sign indicates that an immediate danger exists so special precautions are necessary?
   (a) danger
   (b) caution
   (c) safety instruction
   (d) none of the above

2. Which type of sign indicates a possible hazard against which proper precautions should be taken?
   (a) danger
   (b) caution
   (c) safety instruction
   (d) none of the above

3. Which type of sign is white with a green upper panel containing white lettering, with additional wording in black letters on the white background?
   (a) danger
   (b) caution
   (c) safety instruction
   (d) none of the above

4. Tags may be used as a substitute for accident prevention signs.
   ____ True ____ False

5. A caution tag is what color scheme?
   (a) a white tag with red ink
   (b) a white tag with red and blank ink
   (c) a yellow tag with black ink
   (d) none of the above

Module 28 - Hazard Communication, Part 2

1. The Hazard Communication Standard is based on the three complementary parts of a comprehensive hazard communication program—labels, safety data sheets, and training.

    ____ True   ____ False

2. The best way to prepare a comprehensive list of hazardous chemicals in the workplace is to do what?

    (a) ask another employer in a similar industry

    (b) ask the person in charge of ordering chemicals

    (c) survey the workplace

    (d) list all of the chemicals you can think of

3. Safety data sheets must be ________________ employees when they are in their work areas during their workshifts.

    (a) readily available to

    (b) handed to

    (c) discussed with

    (d) none of the above

4. When must employees working with hazardous chemicals be provided information and training on those chemicals?

    (a) prior to initial assignment to work with a hazardous chemical

    (b) whenever the hazard changes

    (c) at a convenient time

    (d) both (a) and (b)

5. The role of a/an ____________________ is to provide detailed information on each hazardous chemical, including its potential hazardous effects, its physical and chemical characteristics, and recommendations for appropriate protective measures.

    (a) label

    (b) safety data sheet

    (c) chemical information officer

    (d) pictogram

Module 29 - Hazardous Waste Operations and Emergency Response (HAZWOPER)

1. Which of the following is **not** covered by the HAZWOPER Standard?
   (a) construction operations using hazardous chemicals
   (b) clean-up operations
   (c) treatment, storage, and disposal facility operations
   (d) emergency response operations

2. When must the written emergency response plan be developed and implemented?
   (a) as soon as a company is formed
   (b) before emergency response operations commence
   (c) immediately after the first emergency response operation concludes
   (d) none of the above

3. Employers that will evacuate employees from the danger area when an emergency occurs, and that do not permit any employees to assist in handling the emergency, are exempt from the requirement to have a written emergency response plan so long as they have a proper emergency action plan.

   ____ True ____ False

4. Which of the following is **not** required to be included in the safety and health program for employees involved in hazardous waste operations?
   (a) a record of all previous emergency responses
   (b) a training program
   (c) a medical surveillance program
   (d) a comprehensive workplan that includes anticipated clean-up activities and normal operation procedures

5. Which type of emergency responder would respond to releases or potential releases for the purpose of stopping the release?
   (a) first responder awareness level
   (b) first responder operations level
   (c) hazardous materials technician
   (d) on-scene incident commander

Module 30 - Ergonomics

1. Ergonomics is enforced by OSHA under the ________________, which requires employers to furnish a workplace free from recognized hazards that could cause death or serious injury to employees.
   (a) Ergonomics Standard
   (b) General Duty Clause
   (c) Ergonomics Standard for general industry
   (d) none of the above
2. Which of the following is **not** an example of a musculoskeletal disorder (MSD)?
   (a) a broken arm
   (b) carpal tunnel syndrome
   (c) swelling of the tendons
   (d) low back pain
3. Which of the following is a factor that could lead to MSDs?
   (a) awkward postures
   (b) forceful lifting, pushing, or pulling
   (c) prolonged repetitive motion
   (d) contact stress or vibration
   (e) all of the above
4. Approximately _________ of all injuries suffered by construction workers are back injuries.
   (a) 5%
   (b) 15%
   (c) 25%
   (d) 50%
5. When you lift materials, you should:
   (a) bend at your knees, not at your waist
   (b) lift with your legs, not your back
   (c) keep the weight you are lifting as far from your body as possible
   (d) both (a) and (b)
   (e) both (b) and (c)

Module 31 - Safety and Health Program

1. OSHA's recommended practices for developing a safety and health program emphasize a proactive approach to managing workplace safety and health.

   ____True ____False

2. The main goal of a safety and health program is to prevent jobsite injuries, illnesses, and deaths, as well as the suffering and financial hardship such events can cause workers, their families, and their employers. Which of the following is not another benefit of a safety and health program?

   (a) It improves product, process, and service quality.

   (b) It betters workplace morale.

   (c) It improves employee recruiting and retention.

   (d) It creates a more favorable image and reputation among customers, suppliers, and the community.

   (e) None of the above, as they are all benefits of a safety and health program.

3. Which of the following is **not** one of the seven core elements of an effective occupational safety and health program?

   (a) Worker Participation

   (b) Punishment for Accidents and Injuries

   (c) Hazard Identification and Assessment

   (d) Education and Training

4. Providing for investigation of "actual" incidents and "near miss" incidents so causes and means for their prevention are identified would fall under which core element of an effective occupational safety and health program?

   (a) Hazard Identification and Assessment

   (b) Hazard Prevention and Control

   (c) Education and Training

   (d) Management Leadership

5. How often should the safety and health program be evaluated to ensure that it is implemented and is operating?

   (a) initially

   (b) initially and at least monthly

   (c) initially and at least annually

   (d) monthly or annually as decided by the employer

Module 32 - Industrial Hygiene

1. _________________ is the art and science of anticipating, recognizing, evaluating, and controlling occupational and environment health hazards in the workplace.
   (a) Ergonomics
   (b) Industrial hygiene
   (c) Ventilation
   (d) Industrial science
2. Which of the following is **not** something OSHA's industrial hygienists do?
   (a) spot causal conditions and help eliminate or control causal conditions through appropriate measures
   (b) play a major role in developing and issuing OSHA standards to protect workers from health hazards associated with toxic chemicals, biological hazards, and harmful physical agents
   (c) write citations to and collect fines from employers
   (d) provide technical assistance and support to the agency's national and regional offices
3. Which of the following is **not** true about radiation?
   (a) Distance is a valuable tool in controlling exposure to both ionizing and non-ionizing radiation.
   (b) The shorter the time of exposure, the smaller the radiation danger.
   (c) Shielding is a way to protect against radiation.
   (d) The greater the protective mass between a radioactive source and the worker, the greater the radiation exposure.
4. Which of the following generally does **not** cause heat stress?
   (a) high air temperature or humidity
   (b) radiant heat sources
   (c) direct physical contact with hot objects
   (d) strenuous physical activities
   (e) none of the above, as they can all lead to heat stress
5. The five major types of engineering controls used to reduce heat stress in hot work environments are air cooling, fans, shielding, insulation, and ice baths.

   ____True ____False

Module 33 - Recordkeeping and Reporting

1. Which of the following must be reported to OSHA within eight hours?
   (a) an employee dies in a motor vehicle accident on a public street that is not undergoing construction
   (b) two employees are sent to the emergency room but are not admitted for in-patient care after a beaker breaks and exposes them to a hazardous substance
   (c) four employees are hospitalized after a commercial train accident
   (d) an employee dies after the forklift she was using to move pallets in her employer's warehouse tips over on her

2. Employers in which of the following industries are **not** required to maintain a 300 Log because they are partially exempt as a "low hazard" industry?
   (a) liquor stores
   (b) child day care services
   (c) funeral service and crematories
   (d) all of the above

3. Good Company, Inc. serves free lunches to all of its employees; Sally Eater, an employee of Good Company, Inc., becomes ill after eating her employer-provided lunch. This illness must be recorded on the OSHA recordkeeping forms or their equivalent.

   ____True ____False

4. Under the OSHA regulations, which of the following would be considered first aid?
   (a) using an elastic bandage for an ankle injury
   (b) using a rigid brace for an ankle injury
   (c) using surgical glue to close a wound
   (d) using tweezers to remove a metallic fleck from the eye

5. If a work-related injury leads to loss of consciousness, the injury ______________.
   (a) is never recordable
   (b) is always recordable
   (c) is recordable if the employee remains unconscious for 30 seconds or more
   (d) is recordable if the employee remains unconscious for 15 seconds or more

Module 34 - Access to Employee Exposure and Medical Records

1. Which of the following would **not** be classified as an "employee exposure record"?
   (a) workplace environmental monitoring results
   (b) medical questionnaires
   (c) biological monitoring results
   (d) safety data sheets
2. Which of the following types of records must be maintained for the duration of employment plus 30 years?
   (a) medical records
   (b) exposure records
   (c) both medical and exposure records
   (d) none of the above
3. Whenever an employee or designated representative requests access to a medical or exposure record, access must be provided within ________________________.
   (a) 1 working day
   (b) 30 calendar days
   (c) 15 working days
   (d) 15 calendar days
4. Employees must be informed at least annually of the existence, location, and availability of any medical or exposure records; the person responsible for maintaining and providing access to records; and each employee's right to access his/her records.
   ____True ____False
5. How long before closing must an employer that is ceasing to do business without a successor employer notify affected current employees of their right of access to their medical and exposure records?
   (a) There is no such requirement.
   (b) at least three months prior to the business closing date
   (c) at least a month prior to the business closing date
   (d) at least a week prior to the business closing date

Module 35 - Inspections, Citations, and Penalties

1. The employer does not have the right to limit an OSHA inspection to only those areas identified by the inspector.

   ____True ____False

2. During the ______________________, the OSHA inspector will discuss possible corrective actions with the employer.

   (a) opening conference

   (b) walkthrough inspection

   (c) closing conference

   (d) none of the above

3. Citations must be posted near the area of each violation for at least ___________ or until the citation is abated, whichever is longer.

   (a) 30 calendar days

   (b) 7 calendar days

   (c) 3 working days

   (d) 1 working day

4. An employer must submit a notice of contest for a citation within ____________ of receipt of the citation or the notice of contest will be considered invalid.

   (a) 3 working days

   (b) 7 calendar days

   (c) 15 working days

   (d) 30 calendar days

5. Which of the following could **not** lead to imprisonment for up to six months per violation?

   (a) a posting violation

   (b) a willful violation that causes the death of an employee

   (c) giving advance notice of an inspection

   (d) knowingly making a false statement on an OSHA 300 Log

Module 36 - Bloodborne Pathogens

1. Pathogenic microorganisms that are present in human blood and other bodily fluid and can infect and cause disease in humans are called ________________.

   (a) bacteria

   (b) bloodborne pathogens

   (c) viruses

   (d) infection

2. OSHA requires employers, including construction employers, to protect any employees having occupational exposure to the hazards of bloodborne pathogens. OSHA's recommendations are to use "universal precautions" and treat everything as infectious.

   ____True ____False

3. Which of the following is **not** an example of an engineering or work practice control that could be used to minimize employee exposure to bloodborne pathogens?

   (a) minimization of the splashing or spraying of blood

   (b) recording all injuries involving bloodborne pathogens on the proper OSHA form

   (c) handwashing

   (d) prevention of needlesticks

4. The employer must make the ________________ vaccine available to all employees who have occupational exposure to bloodborne pathogens.

   (a) hepatitis B

   (b) hepatitis C

   (c) HIV

   (d) all of the above

5. While performing her job, Sally suffers a deep cut on a broken piece of glass. Her cut requires stitches. While cleaning up the glass involved with Sally's injury, John suffers a minor cut on a piece of glass contaminated with Sally's blood. His injury requires only first aid. Do the injuries need to be recorded on the OSHA injury and illness logs?

   (a) Only Sally's injury must be recorded.

   (b) Only John's injury must be recorded.

   (c) Both injuries must be recorded—Sally's because it required medical attention beyond first aid, and John's because it involved a cut from a sharp object contaminated with another person's blood.

   (d) Both injuries must be recorded—Sally's because it required medical attention beyond first aid, and John's because it involved first aid.

Module 37 - Open Forum - No Review Quiz

ESSENTIALS OF
SAFETY

SAFETY DATA SHEET
Mighty Bond

SECTION 1 - IDENTIFICATION

PRODUCT NAME: Mighty Bond
CHEMICAL FAMILY NAME: Not Available
PRODUCT USE: Adhesive.
SUPPLIER/MANUFACTURER'S NAME: Half-Life Chemical
ADDRESS: 1907 Black Mesa Drive, Colfax, NM 87710. USA
EMERGENCY PHONE: TOLL-FREE in USA: 1-555-234-5678
BUSINESS PHONE: 1-555-876-5432
WEB SITE: www.halflifechemical.com

SECTION 2 - HAZARDS IDENTIFICATION

OSHA Hazard Classification(s):
Flammable liquid, Category 2
Eye irritation, Category 2A

Signal Word: **Warning!**

Hazard Statement(s):
Highly flammable liquid and vapor.
Causes severe eye irritation.

US OSHA HAZARD SYMBOL

US DOT SYMBOLS

Precautionary Statement(s):
Keep container tightly closed.
Keep away from heat/sparks/open flame. – No smoking.
Wear protective gloves and eye/face protection.
Ground/Bond container and receiving equipment.
Use explosion-proof electrical/ventilating/lighting/
equipment.
Take precautionary measures against static discharge.
Use only non-sparking tools.
Store in cool/well-ventilated place.
Avoid release to the environment.

SECTION 3 - COMPOSITION / INFORMATION ON INGREDIENTS

Chemical identity: Component A 70-80%
Common name: Solvent A
Numbers of identity: CAS-Nr.:111111-11-1
Impurities: None

Chemical identity: Component C 20-25%
Common name: Not applicable
Numbers of identity: CAS-Nr.: 44444-44-4
Impurities: None
US OSHA HAZARD SYMBOL US DOT SYMBOLS
Non-Regulated

Trade Secret: The specific chemical identity and exact percentage (concentration) of composition has been withheld as a trade secret.

NOTE: All WHMIS required information is included in appropriate sections based on the ANSI Z400. 1-2004 format. This product has been classified in accordance with the hazard criteria of the CPR and the SDS contains all the information required by the CPR, EU Directives and the Japanese Industrial Standard *JIS Z 7250: 2000.*

SECTION 12 - ECOLOGICAL INFORMATION

ALL WORK PRACTICES MUST BE AIMED AT ELIMINATING ENVIRONMENTAL CONTAMINATION.
EFFECT OF CHEMICAL ON AQUATIC LIFE: No evidence is currently available on this product's effects on aquatic life.
DEGRADATION & PERSISTENCE: The total of the organic components contained in the product is not classified as "readily biodegradable" (OECD-301 A-F). However, this product is expected to be inherently biodegradable.
BIOACCUMULATIVE POTENTIAL: Low
MOVEMENT IN SOIL: Accidental spillage may lead to penetration in the soil and groundwater. However, there is no evidence that this would cause adverse ecological effects.
EFFECT OF MATERIAL ON PLANTS or ANIMALS: No evidence is currently available on this product's effects on plants or animals.

SECTION 13 - DISPOSAL CONSIDERATIONS

PREPARING WASTES FOR DISPOSAL: Waste disposal must be in accordance with appropriate Federal, State, and local regulations, those of Canada, Australia, EU Member States and Japan.

SECTION 14 - TRANSPORTATION INFORMATION

UN NUMBER: UN 1760
PROPER SHIPPING NAME: Corrosive Liquid, n.o.s. (contains Potassium Hydroxide and Sodium Hypochlorite)
HAZARD CLASS NUMBER AND DESCRIPTION: 8
PACKING GROUP: II
DOT LABEL(S) REQUIRED: Corrosive
NORTH AMERICAN EMERGENCY RESPONSE GUIDEBOOK NUMBER (2004): 154
MARINE POLLUTANT: None of the ingredients are classified by the DOT as a Marine Pollutant (as defined by 49 CFR 172.101, Appendix B)

SECTION 15 - REGULATORY INFORMATION

UNITED STATES REGULATIONS
SARA REPORTING REQUIREMENTS: This product is not subject to the reporting requirements of Sections 302, 304 and 313 of Title III of the Superfund Amendments and Reauthorization Act., as follows: None
TSCA: All components in this product are listed on the US Toxic Substances Control Act (TSCA) inventory of chemicals.
Acute Health: Yes
Chronic Health: No
Fire: No
Reactivity: No
U.S. SARA THRESHOLD PLANNING QUANTITY: There are no specific Threshold Planning Quantities for this product. The default Federal MSDS submission and inventory requirement filing threshold of 10,000 lb (4,540 kg) may apply, per 40 CFR 370.20.
U.S. CERCLA REPORTABLE QUANTITY (RQ): None
CALIFORNIA SAFE DRINKING WATER AND TOXIC ENFORCEMENT ACT (PROPOSITION 65): None of the ingredients are on the California Proposition 65 lists.
CANADIAN REGULATIONS
CANADIAN DSL/NDSL INVENTORY STATUS: All of the components of this product are on the DSL Inventory
CANADIAN ENVIRONMENTAL PROTECTION ACT (CEPA) PRIORITIES SUBSTANCES LISTS: No component of this product is on the CEPA First Priorities Substance Lists.
CANADIAN WHMIS CLASSIFICATION and SYMBOLS: This product is categorized as a Controlled Product, Hazard Class E as per the Controlled Product Regulations
EUROPEAN ECONOMIC COMMUNITY INFORMATION:
EU LABELING AND CLASSIFICATION:
Classification of the mixture according to Regulation (EC) No1272/2008. See section 2 for details.

SECTION 16 - OTHER INFORMATION

PREPARED BY: Gordon Freeman, Half-Life Chemical
DATE OF PREPARATION: May 2011
DATE OF LAST REVISION: June 2012

Disclaimer: To the best of Half-Life Chemical's knowledge, the information contained herein is reliable and accurate as of this date; however, accuracy, suitability or completeness is not guaranteed and no warranties of any type either express or implied are provided.
The information contained herein relates only to this specific product.

SECTION 8 - EXPOSURE CONTROLS - PERSONAL PROTECTION CONT.

Currently, international exposure limits are not established for the components of this product. Please check with competent authority in each country for the most recent limits in place.

VENTILATION AND ENGINEERING CONTROLS: Use with adequate ventilation to ensure exposure levels are maintained below the limits provided below. Use local exhaust ventilation to control airborne dust. Ensure eyewash/safety shower stations are available near areas where this product is used.

The following information on appropriate Personal Protective Equipment is provided to assist employers in complying with OSHA regulations found in 29 CFR Subpart I (beginning at 1910.132) or equivalent standard of Canada, or standards of EU member states (including EN 149 for respiratory PPE, and EN 166 for face/eye protection), and those of Japan. Please reference applicable regulations and standards for relevant details.

RESPIRATORY PROTECTION: Maintain airborne contaminant concentrations below guidelines listed above, if applicable. If necessary, use only respiratory protection authorized in the U.S. Federal OSHA Respiratory Protection Standard (29 CFR 1910.134), equivalent U.S. State standards, Canadian CSA Standard Z94.4-93, the European Standard EN149, or EU member states.
EYE PROTECTION: Safety glasses. If necessary, refer to U.S. OSHA 29 CFR 1910.133 or appropriate Canadian Standards.
HAND PROTECTION: Use chemical resistant gloves to prevent skin contact. If necessary, refer to U.S. OSHA 29 CFR 1910.138 or appropriate Standards of Canada.
BODY PROTECTION: Use body protection appropriate to prevent contact (e.g. lab coat, overalls). If necessary, refer to appropriate Standards of Canada, or appropriate Standards of the EU, Australian Standards, or relevant Japanese Standards.

SECTION 9 - PHYSICAL AND CHEMICAL PROPERTIES

APPEARANCE: Clear colorless liquid
ODOR: Slight chlorine odor
ODOR THRESHOLD (PPM): Mild
pH: Not applicable
FREEZING POINT (C°): Not available
BOILING POINT (C°): 132.8°F (56°C)
FLASH POINT: 22°C DIN 51755
EVAPORATION RATE: Not Available
FLAMMABILITY: Not Applicable
EXPLOSIVE LIMITS: lower limit = 1.4 Vol%; upper limit 13.0 Vol%
VAPOR PRESSURE (mmHg): 240 mbar (highest partial vapor pressure) at 20°C
VAPOR DENSITY (AIR=1): Not Available
RELATIVE DENSITY: 0.89 g/cm3 at 20°C
SOLUBILITY(IES): Partially soluble in water at 20°C
PARTITION COEFFICIENT: N-OCTANOL/WATER: Log Kow = 3.3
AUTO-IGNITION TEMPERATURE: Not Available
DECOMPOSITION TEMPERATURE: Not available
VISCOSITY: Low

SECTION 10 - STABILITY AND REACTIVITY

REACTIVITY: Not Available
STABILITY: Product is stable
HAZARDOUS POLYMERIZATION: Will not occur.
CONDITIONS TO AVOID: Heat, sparks, flame and build up of static electricity.
MATERIALS WITH WHICH SUBSTANCE IS INCOMPATIBLE: Halogens, strong acids, alkalies and oxidizers.
DECOMPOSITION PRODUCTS: None are known.

SECTION 11 - TOXICOLOGICAL INFORMATION

IRRITANCY OF PRODUCT: Contact with this product can be irritating to exposed skin, eyes and respiratory system.
SENSITIZATION OF PRODUCT: This product is not considered a sensitizer.
REPRODUCTIVE TOXICITY INFORMATION: No information concerning the effects of this product and its components on the human reproductive system.
TOXICITY DATA: Toxicity data is not available for mixture:
CAS# 1310-58-3 LD50 Dermal 273 mg/kg
CAS# 1344-09-8 LD50 Dermal 1960 mg/kg
CAS# 7681-52-9 LD50 5800 mg/kg

SUSPECTED CANCER AGENT: None of the ingredients are found on the following lists: FEDERAL OSHA Z LIST, NTP, CAL/OSHA, IARC and therefore is not considered to be, nor suspected to be a cancer-causing agent by these agencies.

SAMPLE
Mighty Bond

SECTION 4 - FIRST-AID MEASURES

EYE CONTACT: Hold eyelids apart and flush eyes with plenty of water for at least 15 minutes. Get medical attention.
SKIN CONTACT: Wash the contaminated area with soap and water. Remove contaminated clothing and wash before reuse. If irritation develops, get medical attention.
INHALATION: Remove person to fresh air. If respiratory irritation, dizziness, nausea, or unconsciousness occurs, seek immediate medical assistance. If breathing has stopped, give artificial respiration.
INGESTION: If swallowed, do NOT induce vomiting. Seek immediate medical attention.

SECTION 5 - FIRE-FIGHTING MEASURES

FIRE EXTINGUISHING MATERIALS: As appropriate for surrounding fire. Carbon dioxide, foam, dry chemical, halon, or water fog. Do not use graphite-based extinguishing materials.
FLASH POINT: >200°F
AUTOIGNITION TEMPERATURE: Not Applicable
FLAMMABLE LIMITS: Lower (LEL): Not Applicable Upper (UEL): Not Applicable
UNUSUAL FIRE AND EXPLOSION HAZARDS: This product is flammable, but has no known explosion hazards.

SECTION 6 - ACCIDENTAL RELEASE MEASURES

PERSONAL PRECAUTIONS:
Depending on extent of release, consider the need for firefighters/emergency responders with adequate personal protective equipment for cleaning up.

Do not eat, drink or smoke while cleaning up. Use a self-contained respirator, a mask with filter (type A class 3) or a filtering mask (e.g., EN 405). Wear protective clothing, safety glasses and impervious gloves (e.g., neoprene gloves). Ensure adequate ventilation. Avoid all sources of ignition, hot surfaces and open flames (see also Section 7).

SPILLS: Eliminate all ignition sources. Runoff may create fire or explosion hazard in sewer system. Absorb on fire retardant, liquid-absorbing material (treated sawdust, diatomaceous earth, sand.) Shovel up and dispose of at an appropriate waste disposal facility in accordance with current applicable laws and regulations, and product characteristics at time of disposal (see also Section 13).

SECTION 7 - HANDLING and STORAGE

WORK PRACTICES AND HYGIENE PRACTICES: Avoid contact with eyes. Avoid prolonged repeated skin contact and breathing mist/vapors. Use in well ventilated area away from all ignition sources. Switch off all electrical devices such as parabolic heaters, hotplates, storage heaters etc. in good time for them to have cooled down before commencing work. Do not smoke; do not weld. Do not empty waste into sanitary drains. Take measures to prevent the build up of electrostatic charge.

STORAGE AND HANDLING PRACTICES: Containers of this product must be properly labeled. Store containers in a cool, dry location. Keep container tightly closed when not in use. Store away from strong oxidizers.

SECTION 8 - EXPOSURE CONTROLS - PERSONAL PROTECTION

INFORMATION ON THE SYSTEM DESIGN:
Draw off vapours directly at the point of generation and exhaust from the work area. In the case of regular work, provide bench-mounted extraction equipment.

EXPOSURE LIMITS/GUIDELINES:

Chemical Name	CAS#	ACGIH TWA	OSHA TWA	SWA
Water	7732-18-5	Not Listed	Not Listed	Not Listed
Potassium Hydroxide	1310-58-3	2 mg/m³	2 mg/m³	2 mg/m³
Tetrapotassium Pyrophosphate	7320-34-5	Not Listed	Not Listed	Not Listed
Sodium Silicate	1344-09-8	Not Listed	Not Listed	Not Listed
Sodium Hypochlorite	7681-52-9	Not Listed	Not Listed	Not Listed
Sodium Gluconate	527-07-1	Not Listed	Not Listed	Not Listed

6 88550 00550 0

SAFETY DATA SHEET

Hydrogen Peroxide

Prepared to GHS Rev04(2011): U.S. OSHA, CMA, ANSI, Canadian WHMIS Standards, Australian WorkSafe, Japanese Industrial Standard JIS Z 7250:2000, and European Directives

Printing date 01/10/2013 Reviewed on 01/10/2013

1 Product and Company Identification

Product identifier
Trade name: Hydrogen Peroxide
CAS Number:
7722-84-1
Details of the supplier of the safety data sheet
Manufacturer/Supplier:
Chemtex Laboratories
528 4th Street, Fakeville, IA 52801
info@chemtexlaboratories.com
555-123-4567 or 555-987-6543 for product information
Emergency telephone number: 1-234-567-8910

2 Hazards identification

Classification of the substance or mixture

GHS03 Flame over circle
Ox. Liq. 2 H272 May intensify fire; oxidizer.

GHS05 Corrosion
Eye Dam. 1 H318 Causes serious eye damage.

GHS07
Acute Tox. 4 H302 Harmful if swallowed.
Skin Irrit. 2 H315 Causes skin irritation.
STOT SE 3 H335-H336 May cause respiratory irritation. May cause drowsiness or dizziness.
Label elements
GHS label elements: The product is classified and labelled according to the Globally Harmonized System (GHS).
Hazard pictograms

GHS03 GHS05 GHS07

Signal word: Danger
Hazard-determining components of labelling:
hydrogen peroxide solution
Hazard statements
H272 May intensify fire; oxidizer.
H302 Harmful if swallowed.
H315 Causes skin irritation.
H318 Causes serious eye damage.
H335-H336 May cause respiratory irritation. May cause drowsiness or dizziness.
Precautionary statements
P101 If medical advice is needed, have product container or label at hand.
P102 Keep out of reach of children.
P103 Read label before use.
P221 Take any precaution to avoid mixing with combustibles.
P210 Keep away from heat/sparks/open flames/hot surfaces. No smoking.
P305+P351+P338 IF IN EYES: Rinse cautiously with water for several minutes. Remove contact lenses, if present and easy to do. Continue rinsing.
P310 Immediately call a POISON CENTER or doctor/physician.
P405 Store locked up.
P501 Dispose of contents/container in accordance with local/regional/national/international regulations.
Classification system:
NFPA ratings (scale 0 - 4)

Health = 3
Fire = 0
Reactivity = 1

HMIS-ratings (scale 0 - 4)

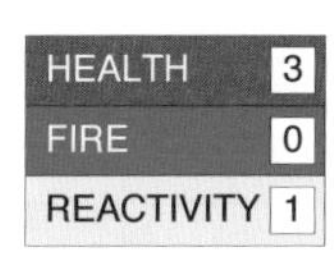

Health = 3
Fire = 0
Reactivity = 1

12 Ecological information

Toxicity
Aquatic toxicity: No further relevant information available.
Persistence and degradability: No further relevant information available.
Behavior in environmental systems:
Bioaccumulative potential: No further relevant information available.
Mobility in soil: No further relevant information available.
Additional ecological information:
General notes:
Water hazard class 1 (Self-assessment): slightly hazardous for water
Do not allow undiluted product or large quantities of it to reach ground water, water course or sewage system.
Results of PBT and vPvB assessment:
PBT: Not applicable.
vPvB: Not applicable.
Other adverse effects: No further relevant information available.

13 Disposal considerations

Waste treatment methods
Recommendation:
Must not be disposed of together with household garbage. Do not allow product to reach sewage system.
Uncleaned packagings:
Recommendation: Disposal must be made according to official regulations.

14 Transport information

UN-Number: None
DOT, ADR, IMDG, IATA UN2014
UN proper shipping name: None
DOT, IMDG, IATA HYDROGEN PEROXIDE, AQUEOUS SOLUTION
ADR 2014 HYDROGEN PEROXIDE, AQUEOUS SOLUTION
Transport hazard class(es):
DOT
Class 5.1 Oxidizing substances.
Label 5.1, 8
ADR
Class 5.1 Oxidizing substances
Label 5.1, 8
IMDG, IATA
Class 5.1 Oxidizing substances.
Label 8
Packing group
DOT, ADR, IMDG, IATA II

Environmental hazards:
Marine pollutant: No
Special precautions for user Warning: Oxidizing substances
Danger code (Kemler): 80
EMS Number: F-A,S-B
Segregation groups Peroxides
Transport in bulk according to Annex II of
MARPOL73/78 and the IBC Code Not applicable.
Transport/Additional information:
DOT
Remarks: Hydrogen Peroxide, Aqueous Solution with more than 20% & <40%
UN "Model Regulation": UN2014, HYDROGEN PEROXIDE, AQUEOUS SOLUTION, 5.1,8,II

15 Regulatory information

Safety, health and environmental regulations/legislation specific for the substance or mixture
Sara: None Listed
Section 355 (extremely hazardous substances):
7722-84-1 hydrogen peroxide solution
Section 313 (Specific toxic chemical listings):
None of the ingredients is listed.
TSCA (Toxic Substances Control Act):
All ingredients are listed.
Proposition 65
Chemicals known to cause cancer:
None of the ingredients is listed.
Chemicals known to cause reproductive toxicity for females:
None of the ingredients is listed.
Chemicals known to cause reproductive toxicity for males:
None of the ingredients is listed.
Chemicals known to cause developmental toxicity:
None of the ingredients is listed.

Carcinogenic categories
EPA (Environmental Protection Agency)
None of the ingredients is listed.
TLV (Threshold Limit Value established by ACGIH)
7722-84-1 hydrogen peroxide solution A3
NIOSH-Ca (National Institute for Occupational Safety and Health)
None of the ingredients is listed.
OSHA-Ca (Occupational Safety & Health Administration)
None of the ingredients is listed.
GHS label elements The product is classified and labelled according to the Globally Harmonized System (GHS).

16 Other information

This information is based on our present knowledge. However, this shall not constitute a guarantee for any specific product features and shall not establish a legally valid contractual relationship.

8 Exposure controls/personal protection

Additional information about design of technical systems: No further data; see item 7.

Control parameters

Components with limit values that require monitoring at the workplace:

7722-84-1 hydrogen peroxide solution

PEL 1.4 mg/m³, 1 ppm

REL 1.4 mg/m³, 1 ppm

TLV 1.4 mg/m³, 1 ppm

Additional information: The lists that were valid during the creation were used as basis.

Exposure controls

Personal protective equipment:

General protective and hygienic measures:

Keep away from foodstuffs, beverages and feed.

Immediately remove all soiled and contaminated clothing.

Wash hands before breaks and at the end of work.

Avoid contact with the skin.

Avoid contact with the eyes and skin.

Breathing equipment:

In case of brief exposure or low pollution use respiratory filter device.

In case of intensive or longer exposure use respiratory protective device that is independent of circulating air.

Protection of hands:

Protective gloves

The glove material has to be impermeable and resistant to the product/ the substance/ the preparation.

Due to missing tests no recommendation to the glove material can be given for the product/ the preparation/ the chemical mixture.

Selection of the glove material on consideration of the penetration times, rates of diffusion and the degradation

Material of gloves

The selection of the suitable gloves does not only depend on the material, but also on further marks of quality and varies from manufacturer to manufacturer. As the product is a preparation of several substances, the resistance of the glove material can not be calculated in advance and has therefore to be checked prior to the application.

Penetration time of glove material

The exact break through time has to be found out by the manufacturer of the protective gloves and has to be observed.

Eye protection: Tightly sealed goggles

9 Physical and chemical properties

Information on basic physical and chemical properties

General Information

Appearance:

Form: Liquid

Color: Colorless

Odor: Odorless

Odor threshold: Not determined.

pH-value at 20 °C (68 °F): 3.7

Change in condition

Melting point/Melting range: ~32 °C

Boiling point/Boiling range: 103 °C (217 °F)

Flash point: Not applicable.

Flammability (solid, gaseous): Not applicable.

Ignition temperature:

Decomposition temperature: Not determined.

Auto igniting: Product is not self-igniting.

Danger of explosion: Product does not present an explosion hazard.

Explosion limits:

Lower: Not determined.

Upper: Not determined.

Vapor pressure at 20 °C (68 °F): 23 hPa (17 mm Hg)

Density at 20 °C (68 °F): 1.15 g/cm³ (9.597 lbs/gal)

Relative density: Not determined.

Vapor density: Not determined.

Evaporation rate: Not determined.

Solubility in / Miscibility with

Water: Not miscible or difficult to mix.

Partition coefficient (n-octanol/water): Not determined.

Viscosity:

Dynamic: Not determined.

Kinematic: Not determined.

Solvent content:

Organic solvents: 0.0 %

Water: 60.0 %

Other information No further relevant information available.

10 Stability and reactivity

Reactivity

Chemical stability

Thermal decomposition / conditions to be avoided: No decomposition if used according to specifications.

Possibility of hazardous reactions: No dangerous reactions known.

Conditions to avoid: No further relevant information available.

Incompatible materials: No further relevant information available.

Hazardous decomposition products: No dangerous decomposition products known.

11 Toxicological information

Information on toxicological effects

Acute toxicity:

Primary irritant effect:

on the skin: Irritant to skin and mucous membranes.

on the eye: Strong irritant with the danger of severe eye injury.

Sensitization: No sensitizing effects known.

Additional toxicological information:

The product shows the following dangers according to internally approved calculation methods for preparations:

Harmful

Irritant

Carcinogenic categories:

IARC (International Agency for Research on Cancer)

7722-84-1 hydrogen peroxide solution 3

NTP (National Toxicology Program)

None of the ingredients are listed.

Other hazards
Results of PBT and vPvB assessment
PBT: Not applicable.
vPvB: Not applicable.

3 Composition/information on ingredients

7732-18-5 water, distilled, conductivity or of similar purity 10-60%
Chemical characterization: Mixtures
Description: Mixture of the substances listed below with nonhazardous additions.
Dangerous components:
7722-84-1 hydrogen peroxide solution
Ox. Liq. 1, H271; Skin Corr. 1A, H314; Acute Tox. 4, H302; Acute Tox. 4, H332

4 First-aid measures

Description of first-aid measures
General information:
Symptoms of poisoning may even occur after several hours; therefore medical observation for at least 48 hours after the accident.
After inhalation: In case of unconsciousness place patient stably in side position for transportation.
After skin contact: Immediately wash with water and soap and rinse thoroughly.
After eye contact: Rinse opened eye for several minutes under running water. Then consult a doctor.
After swallowing: Immediately call a doctor.
Information for doctor:
Most important symptoms and effects, both acute and delayed. No further relevant information available.
Indication of any immediate medical attention and special treatment needed.
No further relevant information available.

5 Firefighting measures

Extinguishing media
Suitable extinguishing agents:
CO2, extinguishing powder or water spray. Fight larger fires with water spray or alcohol resistant foam.
Special hazards arising from the substance or mixture. No further relevant information available.
Advice for firefighters:
Protective equipment: No special measures required.

6 Accidental release measures

Personal precautions, protective equipment and emergency procedures: Not required.
Environmental precautions: Do not allow to enter sewers/ surface or ground water.
Methods and material for containment and cleaning up:
Absorb with liquid-binding material (sand, diatomite, acid binders, universal binders, sawdust).
Dispose contaminated material as waste according to item 13.
Ensure adequate ventilation.
Reference to other sections
See Section 7 for information on safe handling.
See Section 8 for information on personal protection equipment.
See Section 13 for disposal information.

7 Handling and storage

Handling:
Precautions for safe handling:
Ensure good ventilation/exhaustion at the workplace.
Prevent formation of aerosols.
Information about protection against explosions and fires: No special measures required.
Conditions for safe storage, including any incompatibilities
Storage:
Requirements to be met by storerooms and receptacles: No special requirements.
Information about storage in one common storage facility: Not required.
Further information about storage conditions: Keep receptacle tightly sealed.
Specific end use(s): No further relevant information available.

SAFETY DATA SHEET
Squeeky Kleen Detergent

SECTION 1 - IDENTIFICATION

PRODUCT NAME: Squeeky Kleen Detergent
CHEMICAL FAMILY NAME: Detergent
PRODUCT USE: Critical - cleaning detergent for laboratory, healthcare and industrial applications.
SUPPLIER/MANUFACTURER'S NAME: Half-Life Chemical
ADDRESS: 1907 Black Mesa Drive, Colfax, NM 87710. USA
EMERGENCY PHONE: TOLL-FREE in USA/Canada: 1-555-234-5678
WEB SITE: www.halflifechemical.com
BUSINESS PHONE: 1-555-876-5432

SECTION 2 - HAZARDS IDENTIFICATION

EMERGENCY OVERVIEW: This product is a clear liquid with a slight chlorine odor. Exposure can be irritating to eyes, respiratory system and skin. This product is corrosive and could cause burns to skin and eyes. It is a non-flammable liquid. The Environmental effects of this product have not been investigated.

OSHA Hazard Classification(s):
Skin Irritant Category 2
Eye Irritant Category 2

Signal Word: Warning!

Hazard Statement(s):
Harmful if swallowed
Causes skin irritation
Causes serious eye irritation

US OSHA HAZARD SYMBOL

US DOT SYMBOLS

Precautionary Statement(s):
Do not breathe dust/fume/gas/mist/vapors/spray.
Wash hands thoroughly after handling.
Use only in well ventilated area.
Wear protective gloves/protective clothing/eye protection/face protection.

EU LABELING AND CLASSIFICATION:
Classification of the substance or mixture according to Regulation (EC) No1272/2008 Annex 1
EC# 231-791-2 This substance is not classified in the Annex I of Directive 67/548/EEC
EC# 215-181-3 Index # 019-002-00-8
EC# 230-785-7 This substance is not classified in the Annex I of Directive 67/548/EEC
EC# 215-687-4 This substance is not classified in the Annex I of Directive 67/548/EEC
EC# 231-668-3 This substance is not classified in the Annex I of Directive 67/548/EEC
EC# 208-407-7 This substance is not classified in the Annex I of Directive 67/548/EEC

SECTION 3 - COMPOSITION / INFORMATION ON INGREDIENTS

HAZARDOUS INGREDIENTS:	CAS #	EINECS #	ICSC #	WT %	HAZARD CLASSIFICATION; RISK PHRASES
Water	7732-18-5	231-791-2	Not Listed	50 – 70%	HAZARD CLASSIFICATION: None RISK PHRASES: None
Potassium Hydroxide	1310-58-3	215-181-3	0357	1 - 10%	HAZARD CLASSIFICATION: [Xn] Harmful, [C] Corrosive RISK PHRASES: R22, R35
Tetrapotassium Pyrophosphate	7320-34-5	231-767-1	1140	3 – 13%	HAZARD CLASSIFICATION: None RISK PHRASES: None
Sodium Silicate	1344-09-8	215-687-4	1137	3 – 8%	HAZARD CLASSIFICATION: None RISK PHRASES: None
Sodium Hypochlorite	7681-52-9	231-668-3	0482	1 – 3%	HAZARD CLASSIFICATION: [C] Corrosive RISK PHRASES: R34
Sodium Gluconate	527-07-1	208-407-7	Not Listed	1 – 3%	HAZARD CLASSIFICATION: None RISK PHRASES: None
Balance of other ingredients are non-hazardous or less than 1% in concentration (or 0.1% for carcinogens, reproductive toxins, or respiratory sensitizers).					

SECTION 13 - DISPOSAL CONSIDERATIONS

PREPARING WASTES FOR DISPOSAL: Waste disposal must be in accordance with appropriate Federal, State, and local regulations, those of Canada, Australia, EU Member States and Japan.

SECTION 14 - TRANSPORTATION INFORMATION

U.N. NUMBER: UN 1760
U.N. DANGEROUS GOODS CLASS: CORROSIVE LIQUID, n.o.s. (contains Potassium Hydroxide and Sodium Hypochlorite)
US DOT; IATA; IMO; ADR:
THIS PRODUCT IS HAZARDOUS AS DEFINED BY 49 CFR 172.101 BY THE U.S. DEPARTMENT OF TRANSPORTATION.
UN IDENTIFICATION NUMBER: UN1760
PROPER SHIPPING NAME: Corrosive Liquid, n.o.s. (contains Potassium Hydroxide and Sodium Hypochlorite)
HAZARD CLASS NUMBER AND DESCRIPTION: Class 8 Corrosive material
PACKING GROUP: PGII
UN IDENTIFICATION NUMBER: UN1760
MARINE POLLUTANT: None of the ingredients are classified by the DOT as a Marine Pollutant (as defined by 49 CFR 172.101, Appendix B)
U.S. DEPARTMENT OF TRANSPORTATION (DOT) SHIPPING REGULATIONS:
This product is classified as dangerous goods, per U.S. DOT regulations, under 49 CFR 172.101.
TRANSPORT CANADA, TRANSPORTATION OF DANGEROUS GOODS REGULATIONS:
This product is classified as Dangerous Goods, per regulations of Transport Canada.
INTERNATIONAL AIR TRANSPORT ASSOCIATION (IATA):
This product is classified as Dangerous Goods, by rules of IATA
INTERNATIONAL MARITIME ORGANIZATION (IMO) DESIGNATION:
This product is classified as Dangerous Goods by the International Maritime Organization.
EUROPEAN AGREEMENT CONCERNING THE INTERNATIONAL CARRIAGE OF DANGEROUS GOODS BY ROAD (ADR):
This product is classified by the United Nations Economic Commission for Europe to be dangerous goods.
DOT LABEL(S) REQUIRED: Corrosive
NORTH AMERICAN EMERGENCY RESPONSE GUIDEBOOK NUMBER (2004): 154

SECTION 15 - REGULATORY INFORMATION

UNITED STATES REGULATIONS
SARA REPORTING REQUIREMENTS: This product is not subject to the reporting requirements of Sections 302, 304 and 313 of Title III of the Superfund Amendments and Reauthorization Act., as follows: None
TSCA: All components in this product are listed on the US Toxic Substances Control Act (TSCA) inventory of chemicals.
Acute Health: Yes Chronic **Health:** No **Fire:** No **Reactivity:** No
U.S. SARA THRESHOLD PLANNING QUANTITY: There are no specific Threshold Planning Quantities for this product. The default Federal SDS submission and inventory requirement filing threshold of 10,000 lb (4,540 kg) may apply, per 40 CFR 370.20.
U.S. CERCLA REPORTABLE QUANTITY (RQ): Potassium Hydroxide CAS# 1310-58-3 - 1,000 Lbs.
CALIFORNIA SAFE DRINKING WATER AND TOXIC ENFORCEMENT ACT (PROPOSITION 65): None of the ingredients are on the California Proposition 65 lists.
CANADIAN REGULATIONS
CANADIAN DSL/NDSL INVENTORY STATUS: All of the components of this product are on the DSL Inventory
CANADIAN ENVIRONMENTAL PROTECTION ACT (CEPA) PRIORITIES SUBSTANCES LISTS: No component of this product is on the CEPA First Priorities Substance Lists.
CANADIAN WHMIS CLASSIFICATION and SYMBOLS: This product is categorized as a Controlled Product, Hazard Class E as per the Controlled Product Regulations
EUROPEAN ECONOMIC COMMUNITY INFORMATION:
EU LABELING AND CLASSIFICATION:
Classification of the mixture according to Regulation (EC) No1272/2008. See section 2 for details.

SECTION 16 - OTHER INFORMATION

PREPARED BY: Gordon Freeman, Half-Life Chemical
DATE OF PREPARATION: May 2011
DATE OF LAST REVISION: June 2012

Disclaimer: To the best of Half-Life Chemical's knowledge, the information contained herein is reliable and accurate as of this date; however, accuracy, suitability or completeness is not guaranteed and no warranties of any type either express or implied are provided. The information contained herein relates only to this specific product.

SAFETY DATA SHEET

Squeeky Kleen Detergent

The following information on appropriate Personal Protective Equipment is provided to assist employers in complying with OSHA regulations found in 29 CFR Subpart I (beginning at 1910.132) or equivalent standard of Canada, or standards of EU member states (including EN 149 for respiratory PPE, and EN 166 for face/eye protection), and those of Japan. Please reference applicable regulations and standards for relevant details.

RESPIRATORY PROTECTION: Maintain airborne contaminant concentrations below guidelines listed above, if applicable. If necessary, use only respiratory protection authorized in the U.S. Federal OSHA Respiratory Protection Standard (29 CFR 1910.134), equivalent U.S. State standards, Canadian CSA Standard Z94.4-93, the European Standard EN149, or EU member states.
EYE PROTECTION: Safety glasses. If necessary, refer to U.S. OSHA 29 CFR 1910.133 or appropriate Canadian Standards.
HAND PROTECTION: Use chemical resistant gloves to prevent skin contact. If necessary, refer to U.S. OSHA 29 CFR 1910.138 or appropriate Standards of Canada.
BODY PROTECTION: Use body protection appropriate to prevent contact (e.g. lab coat, overalls). If necessary, refer to appropriate Standards of Canada, or appropriate Standards of the EU, Australian Standards, or relevant Japanese Standards.

SECTION 9 - PHYSICAL and CHEMICAL PROPERTIES

APPEARANCE: Clear colorless liquid.
ODOR: Slight chlorine odor.
ODOR THRESHOLD (PPM): Not Available
pH: 12.4 (1% solution)
FREEZING POINT (C°): Not Applicable
BOILING POINT (C°): 100°C (212°F)
FLASH POINT: Not Applicable
EVAPORATION RATE (nBuAc = 1): <1
FLAMMABILITY: Not Applicable
UPPER/LOWER FLAMMABILITY OR EXPLOSIVE LIMITS: Not Applicable
VAPOR PRESSURE (mmHg): 17 @ 20°C (68°F)
VAPOR DENSITY (AIR=1): >1
RELATIVE DENSITY: Not Available
SOLUBILITY(IES): Complete
PARTITION COEFFICIENT: N-OCTANOL/WATER: Not Available
AUTO-IGNITION TEMPERATURE: Not Available
DECOMPOSITION TEMPERATURE: 125°C (257°F)
VISCOSITY: 8.95 Pa s

SECTION 10 - STABILITY AND REACTIVITY

REACTIVITY: Not Available
STABILITY: Product is stable
HAZARDOUS POLYMERIZATION: Will not occur.
CONDITIONS TO AVOID: Contact with incompatible materials.
MATERIALS WITH WHICH SUBSTANCE IS INCOMPATIBLE: Strong acids, and strong oxidizing agents.
DECOMPOSITION PRODUCTS: When heated to decomposition, this product produces Oxides of carbon (COx) and Hydrocarbons.

SECTION 11 - TOXICOLOGICAL INFORMATION

IRRITANCY OF PRODUCT: Contact with this product can be irritating to exposed skin, eyes and respiratory system.
SENSITIZATION OF PRODUCT: This product is not considered a sensitizer.
REPRODUCTIVE TOXICITY INFORMATION: No information concerning the effects of this product and its components on the human reproductive system.
TOXICITY DATA: Toxicity data is not available for mixture:
CAS# 1310-58-3 LD50 Oral (Rat) 273 mg/kg
CAS# 1344-09-8 LD50 Oral (Rat) 1960 mg/kg
CAS# 7681-52-9 LD50 Oral (Mouse) 5800 mg/kg

SUSPECTED CANCER AGENT: None of the ingredients are found on the following lists: FEDERAL OSHA Z LIST, NTP, CAL/OSHA, IARC and therefore is not considered to be, nor suspected to be a cancer-causing agent by these agencies.

SECTION 12 - ECOLOGICAL INFORMATION

ALL WORK PRACTICES MUST BE AIMED AT ELIMINATING ENVIRONMENTAL CONTAMINATION.
EFFECT OF CHEMICAL ON AQUATIC LIFE: No evidence is currently available on this product's effects on aquatic life.
Degradation & Persistence: Based on stringent OECD test guidelines, this material cannot be considered as readily biodegradable. However, biodegradation may occur under aerobic conditions (in the presence of oxygen).
Bioaccumulative Potential: Low
Movement in Soil: Potential for mobility in soil is very high (Koc between 0 and 50).
ENVIRONMENTAL STABILITY: No Data available at this time.
EFFECT OF MATERIAL ON PLANTS or ANIMALS: No evidence is currently available on this product's effects on plants or animals.

SECTION 4 - FIRST-AID MEASURES

Individuals who are exposed must be taken for medical attention if any adverse effect occurs. Rescuers should be taken for medical attention, if necessary. Take copy of label and SDS to health professional with contaminated individual.

EYE CONTACT: If product enters the eyes, open eyes while under gentle running water for at least 15 minutes. Seek medical attention.
SKIN CONTACT: Wash skin thoroughly after handling. Seek medical attention if irritation develops and persists. Remove contaminated clothing. Launder before re-use.
INHALATION: If breathing becomes difficult, remove victim to fresh air. If necessary, use artificial respiration to support vital functions. Seek medical attention if breathing difficulty continues.
INGESTION: If product is swallowed, call physician or poison control center for most current information. If professional advice is not available, do not induce vomiting. Never induce vomiting or give diluents (milk or water) to someone who is unconscious, having convulsions, or who cannot swallow. Seek medical advice. Take a copy of the label and/or SDS with the victim to the health professional.
MEDICAL CONDITIONS AGGRAVATED BY EXPOSURE: Pre-existing skin or eye problems may be aggravated by prolonged contact.
RECOMMENDATIONS TO PHYSICIANS: Treat symptoms and reduce over-exposure.
Symptoms / Effects: Exposure to this product may cause irritation of the eyes with redness and swelling. Contact with skin may cause chemical burns and/or irritation with redness and swelling. Inhalation of this product may cause irritation to the respiratory tract. Ingestion may cause gastrointestinal irritation including pain, vomiting or diarrhea.
Chronic: None known

SECTION 5 - FIRE-FIGHTING MEASURES

FIRE EXTINGUISHING MATERIALS: As appropriate for surrounding fire. Carbon dioxide, foam, dry chemical, halon, or water fog. Do not use graphite-based extinguishing materials.
FLASH POINT: Non-Flammable
AUTOIGNITION TEMPERATURE: Not Applicable
FLAMMABLE LIMITS (in air by volume, %): Lower (LEL): Not Applicable Upper (UEL): Not Applicable
UNUSUAL FIRE AND EXPLOSION HAZARDS: Containers exposed to intense heat may rupture.
Explosion Sensitivity to Mechanical Impact: Not Available
Explosion Sensitivity to Static Discharge: Not Available
SPECIAL FIRE-FIGHTING PROCEDURES: Incipient fire responders should wear eye protection. Structural firefighters must wear Self-Contained Breathing Apparatus and full protective equipment. Isolate materials not yet involved in the fire and protect personnel. Move containers from fire area if this can be done without risk; otherwise, cool with carefully applied water spray. If possible, prevent runoff water from entering storm drains, bodies of water, or other environmentally sensitive areas.

SECTION 6 - ACCIDENTAL RELEASE MEASURES

SPILL AND LEAK RESPONSE: Personnel should be trained for spill response operations. Ensure adequate ventilation; use respirator for prolonged exposure. Use gloves, protective clothing, and safety glasses for clean-up.
SPILLS: Contain spill if safe to do so. Prevent entry into drains, sewers, and other waterways. Soak up with an absorbent material and place in an appropriate container for disposal. Dispose of in accordance with applicable Federal, State, and local procedures (see Section 13, Disposal Considerations).

SECTION 7 - HANDLING and STORAGE

WORK PRACTICES AND HYGIENE PRACTICES: As with all chemicals, avoid getting this product ON YOU or IN YOU. Wash thoroughly after handling this product. Do not eat, drink, smoke, or apply cosmetics while handling this product. Avoid breathing vapors/mists/sprays generated by this product. Use in a well-ventilated location. Remove contaminated clothing immediately.
STORAGE AND HANDLING PRACTICES: Containers of this product must be properly labeled. Store containers in a cool, dry location. Keep container tightly closed when not in use. Store away from strong acids or strong oxidizers

SECTION 8 - EXPOSURE CONTROLS / PERSONAL PROTECTION

EXPOSURE LIMITS/GUIDELINES:

Chemical Name	CAS#	ACGIH TWA	OSHA TWA	SWA
Water	7732-18-5	Not Listed	Not Listed	Not Listed
Potassium Hydroxide	1310-58-3	2 mg/m³	2 mg/m³	2 mg/m³
Tetrapotassium Pyrophosphate	7320-34-5	Not Listed	Not Listed	Not Listed
Sodium Silicate	1344-09-8	Not Listed	Not Listed	Not Listed
Sodium Hypochlorite	7681-52-9	Not Listed	Not Listed	Not Listed
Sodium Gluconate	527-07-1	Not Listed	Not Listed	Not Listed

Currently, international exposure limits are not established for the components of this product. Please check with competent authority in each country for the most recent limits in place.

VENTILATION AND ENGINEERING CONTROLS: Use with adequate ventilation to ensure exposure levels are maintained below the limits provided below. Use local exhaust ventilation to control airborne dust. Ensure eyewash/safety shower stations are available near areas where this product is used.

SAFETY DATA SHEET
Sulfuric Acid

Prepared to GHS Rev04(2011): U.S. OSHA, CMA, ANSI, Canadian WHMIS Standards, Australian WorkSafe, Japanese Industrial Standard JIS Z 7250:2000, and European Directives

Printing date 01/10/2013 Reviewed on 01/10/2013

1 Product and Company Identification
Product identifier
Trade name: Sulfuric Acid
CAS Number:
5564-93-9
EC number:
231-555-5
Index number:
016-550-00-8
Details of the supplier of the safety data sheet
Manufacturer/Supplier:
Chemtex Laboratories
528 4th Street, Fakeville, IA 52801
info@chemtexlaboratories.com
555-123-4567 or 555-987-6543 for product information
Emergency telephone number: 1-234-567-8910

2 Hazards identification
Classification of the substance or mixture

GHS05 Corrosion

Skin Corr. 1A H314 Causes severe skin burns and eye damage.
Label elements
GHS label elements
The substance is classified and labelled according to the Globally Harmonized System (GHS).
Hazard pictograms

GHS05

Signal word: Danger
Hazard-determining components of labelling:
sulfuric acid
Hazard statements
H314 Causes severe skin burns and eye damage.
Precautionary statements
P101 If medical advice is needed, have product container or label at hand.
P102 Keep out of reach of children.
P103 Read label before use.
P260 Do not breathe dust/fume/gas/mist/vapors/spray.
P303+P361+P353 IF ON SKIN (or hair): Remove/Take off immediately all contaminated clothing. Rinse skin with water/shower.
P305+P351+P338 IF IN EYES: Rinse cautiously with water for several minutes. Remove contact lenses, if present and easy to do. Continue rinsing.
P310 Immediately call a POISON CENTER or doctor/physician.
P405 Store locked up.
P501 Dispose of contents/container in accordance with local/regional/national/international regulations.
Classification system:
NFPA ratings (scale 0 - 4)

Health = 4
Fire = 0
Reactivity = 2

· HMIS-ratings (scale 0 - 4)

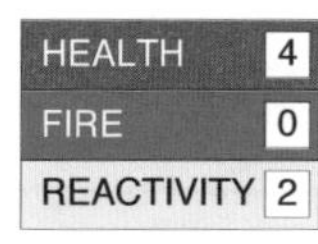

Health = 4
Fire = 0
Reactivity = 2

Other hazards
Results of PBT and vPvB assessment
PBT: Not applicable.
vPvB: Not applicable.

Recommendation: Disposal must be made according to official regulations.
Recommended cleansing agent: Water, if necessary with cleansing agents.

14 Transport information
UN-Number: None
DOT, ADR, IMDG, IATA UN1830
UN proper shipping name: None
DOT, IMDG, IATA SULPHURIC ACID
ADR 1830 SULPHURIC ACID
Transport hazard class(es):
DOT
Class 8 Corrosive substances.
Label 8
ADR, IMDG, IATA
Class 8 Corrosive substances
Label 8
Packing group
DOT, ADR, IMDG, IATA II
Environmental hazards:
Marine pollutant: No
Special precautions for user Warning: Corrosive substances
Danger code (Kemler): 80
EMS Number: F-A,S-B
Segregation groups Acids
Transport in bulk according to Annex II of
MARPOL73/78 and the IBC Code Not applicable.
UN "Model Regulation": UN1830, SULPHURIC ACID, 8, II

15 Regulatory information
Safety, health and environmental regulations/legislation
specific for the substance or mixture
Sara: None Listed
Section 355 (extremely hazardous substances):
Substance is listed.
Section 313 (Specific toxic chemical listings):
Substance is listed.
TSCA (Toxic Substances Control Act):
Substance is listed.
Proposition 65
Chemicals known to cause cancer:
Substance is not listed.
Chemicals known to cause reproductive toxicity for females:
Substance is not listed.
Chemicals known to cause reproductive toxicity for males:
Substance is not listed.
Chemicals known to cause developmental toxicity:
Substance is not listed.
Carcinogenic categories:
EPA (Environmental Protection Agency)
Substance is not listed.
TLV (Threshold Limit Value established by ACGIH)
7664-93-9 sulphuric acid A2
NIOSH-Ca (National Institute for Occupational Safety and Health)
Substance is not listed.
OSHA-Ca (Occupational Safety & Health Administration)
Substance is not listed.
GHS label elements
The substance is classified and labelled according to the Globally
Harmonized System (GHS).

Hazard pictograms
GHS05
Signal word: Danger
Hazard-determining components of labelling:
sulfuric acid
Hazard statements
H314 Causes severe skin burns and eye damage.
Precautionary statements
P101 If medical advice is needed, have product container or label at hand.
P102 Keep out of reach of children.
P103 Read label before use.
P260 Do not breathe dust/fume/gas/mist/vapours/spray.
P303+P361+P353 IF ON SKIN (or hair): Remove/Take off immediately all contaminated clothing. Rinse skin with water/shower.
P305+P351+P338 IF IN EYES: Rinse cautiously with water for several minutes.
Remove contact lenses, if
present and easy to do. Continue rinsing.
P310 Immediately call a POISON CENTER or doctor/physician.
P405 Store locked up.
P501 Dispose of contents/container in accordance with local/regional/national/international regulations.
Chemical safety assessment: A Chemical Safety Assessment has not been carried out.

16 Other information
This information is based on our present knowledge. However, this shall not constitute a guarantee for any specific product features and shall not establish a legally valid contractual relationship.
Abbreviations and acronyms:
ADR: Accord européen sur le transport des marchandises dangereuses par Route (European Agreement concerning the International Carriage of Dangerous Goods by Road)
IMDG: International Maritime Code for Dangerous Goods
DOT: US Department of Transportation
IATA: International Air Transport Association
ACGIH: American Conference of Governmental Industrial Hygienists
EINECS: European Inventory of Existing Commercial Chemical Substances
CAS: Chemical Abstracts Service (division of the American Chemical Society)
NFPA: National Fire Protection Association (USA)
HMIS: Hazardous Materials Identification System (USA)

9 Physical and chemical properties

Information on basic physical and chemical properties
General Information
Appearance:
Form: Fluid
Color: Not determined.
Odor: Mild
Odor threshold: Not determined.
pH-value at 20 °C (68 °F): 0.1
Change in condition
Melting point/Melting range: ~33 °C
Boiling point/Boiling range: 295 °C (563 °F)
Flash point: Not applicable.
Flammability (solid, gaseous): Not applicable.
Ignition temperature:
Decomposition temperature: Not determined.
Auto igniting: Not determined.
Danger of explosion: Product does not present an explosion hazard.
Explosion limits:
Lower: Not determined.
Upper: Not determined.
Vapor pressure at 20 °C (68 °F): <0.01 hPa (<0 mm Hg)
Density at 20 °C (68 °F): 1.84 g/cm^3 (15.355 lbs/gal)
Relative density: Not determined.
Vapour density: Not determined.
Evaporation rate: Not determined.
Solubility in / Miscibility with
Water: Fully miscible.
Partition coefficient (n-octanol/water): Not determined.
Viscosity:
Dynamic: Not determined.
Kinematic: Not determined.
Organic solvents: 0.0 %
Other information: No further relevant information available.

10 Stability and reactivity

Reactivity
Chemical stability
Thermal decomposition / conditions to be avoided: No decomposition if used according to specifications.
Possibility of hazardous reactions: No dangerous reactions known.
Conditions to avoid: No further relevant information available.
Incompatible materials: No further relevant information available.
Hazardous decomposition products: No dangerous decomposition products known.

11 Toxicological information

Information on toxicological effects
Acute toxicity:
Primary irritant effect:
on the skin: Strong caustic effect on skin and mucous membranes.
on the eye: Strong caustic effect.
Sensitization: No sensitizing effects known.
Additional toxicological information:
Swallowing will lead to a strong caustic effect on mouth and throat and to the danger of perforation of esophagus and stomach.
Carcinogenic categories:
IARC (International Agency for Research on Cancer)
5564-93-9 sulphuric acid 1
NTP (National Toxicology Program)
5564-93-9 sulfuric acid K

12 Ecological information

Toxicity
Aquatic toxicity: No further relevant information available.
Persistence and degradability: No further relevant information available.
Behavior in environmental systems:
Bioaccumulative potential: No further relevant information available.
Mobility in soil: No further relevant information available.
Additional ecological information:
General notes:
Water hazard class 1 (Assessment by list): slightly hazardous for water
Do not allow undiluted product or large quantities of it to reach ground water, water course or sewage system.
Must not reach bodies of water or drainage ditch undiluted or unneutralized.
Rinse off bigger amounts into drains or the aquatic environment may lead to decreased pH-values. A low pHvalue harms aquatic organisms. In the dilution of the use-level the pH-value is considerably increased, so that after the use of the product the aqueous waste, emptied into drains, is only low water dangerous.
Results of PBT and vPvB assessment
PBT: Not applicable.
vPvB: Not applicable.
Other adverse effects: No further relevant information available.

13 Disposal considerations

Waste treatment methods
Recommendation:
Must not be disposed of together with household garbage. Do not allow product to reach sewage system.
Uncleaned packagings:

<u>3 Composition/information on ingredients</u>

Chemical characterization: Substances
CAS No. Description
5564-93-9 sulfuric acid

Identification number(s)
EC number: 231-559-5
Index number: 016-020-00-8

<u>4 First-aid measures</u>

Description of first-aid measures
General information: Immediately remove any clothing soiled by the product.
After inhalation: In case of unconsciousness place patient stably in side position for transportation.
After skin contact: Immediately wash with water and soap and rinse thoroughly.
After eye contact: Rinse opened eye for several minutes under running water. Then consult a doctor.
After swallowing: Drink copious amounts of water and provide fresh air. Immediately call a doctor.
Information for doctor:
Most important symptoms and effects, both acute and delayed. No further relevant information available.
Indication of any immediate medical attention and special treatment needed.
No further relevant information available.

<u>5 Firefighting measures</u>

Extinguishing media
Suitable extinguishing agents:
CO2, extinguishing powder or water spray. Fight larger fires with water spray or alcohol resistant foam.
Special hazards arising from the substance or mixture No further relevant information available.
Advice for firefighters:
Protective equipment: No special measures required.

<u>6 Accidental release measures</u>

Personal precautions, protective equipment and emergency procedures
Wear protective equipment. Keep unprotected persons away.
Environmental precautions:
Dilute with plenty of water.
Do not allow to enter sewers/ surface or ground water.

Methods and material for containment and cleaning up:
Absorb with liquid-binding material (sand, diatomite, acid binders, universal binders, sawdust).
Use neutralizing agent.
Dispose contaminated material as waste according to item 13.
Ensure adequate ventilation.

<u>7 Handling and storage</u>

Handling:
Precautions for safe handling. No special precautions are necessary if used correctly.
Information about protection against explosions and fires: No special measures required.
Conditions for safe storage, including any incompatibilities
Storage:
Requirements to be met by storerooms and receptacles: No special requirements.
Information about storage in one common storage facility: Not required.
Further information about storage conditions: Keep receptacle tightly sealed.
Specific end use(s): No further relevant information available.

<u>8 Exposure controls/personal protection</u>

Additional information about design of technical systems:
No further data; see item 7.
Control parameters
Components with limit values that require monitoring at the workplace:
5564-93-9 sulphuric acid

PEL	1 mg/m³
REL	1 mg/m³
TLV	0.2* mg/m³
	*as thoracic fraction

Additional information: The lists that were valid during the creation were used as basis.
Exposure controls
Personal protective equipment:
General protective and hygienic measures:
Keep away from foodstuffs, beverages and feed.
Immediately remove all soiled and contaminated clothing.
Wash hands before breaks and at the end of work.
Avoid contact with the eyes and skin.
Breathing equipment: Not required.

Protection of hands:
Protective gloves
The glove material has to be impermeable and resistant to the product/ the substance/ the preparation.
Due to missing tests no recommendation to the glove material can be given for the product/ the preparation/ the chemical mixture.
Selection of the glove material on consideration of the penetration times, rates of diffusion and the degradation
Material of gloves
The selection of the suitable gloves does not only depend on the material, but also on further marks of quality and varies from manufacturer to manufacturer.
Penetration time of glove material
The exact break through time has to be found out by the manufacturer of the protective gloves and has to be observed.
Eye protection:
Tightly sealed goggles

How to Fill Out the 300 Log

OMB Disclosure Statement: Public reporting burden for this collection of information is estimated to average 14 minutes per response, including time to review the instructions, search and gather the data needed, and complete and review the collection of information. Persons are not required to respond to the collection of information unless it displays a currently valid OMB control number. If you have any comments about these estimates or any other aspects of this data collection, contact: US Department of Labor, OSHA Office of Statistical Analysis, Room N-3644, 200 Constitution Avenue, NW, Washington, DC 20210. Do not send the completed forms to this office.
See www.mancomm.com for updates or revisions.

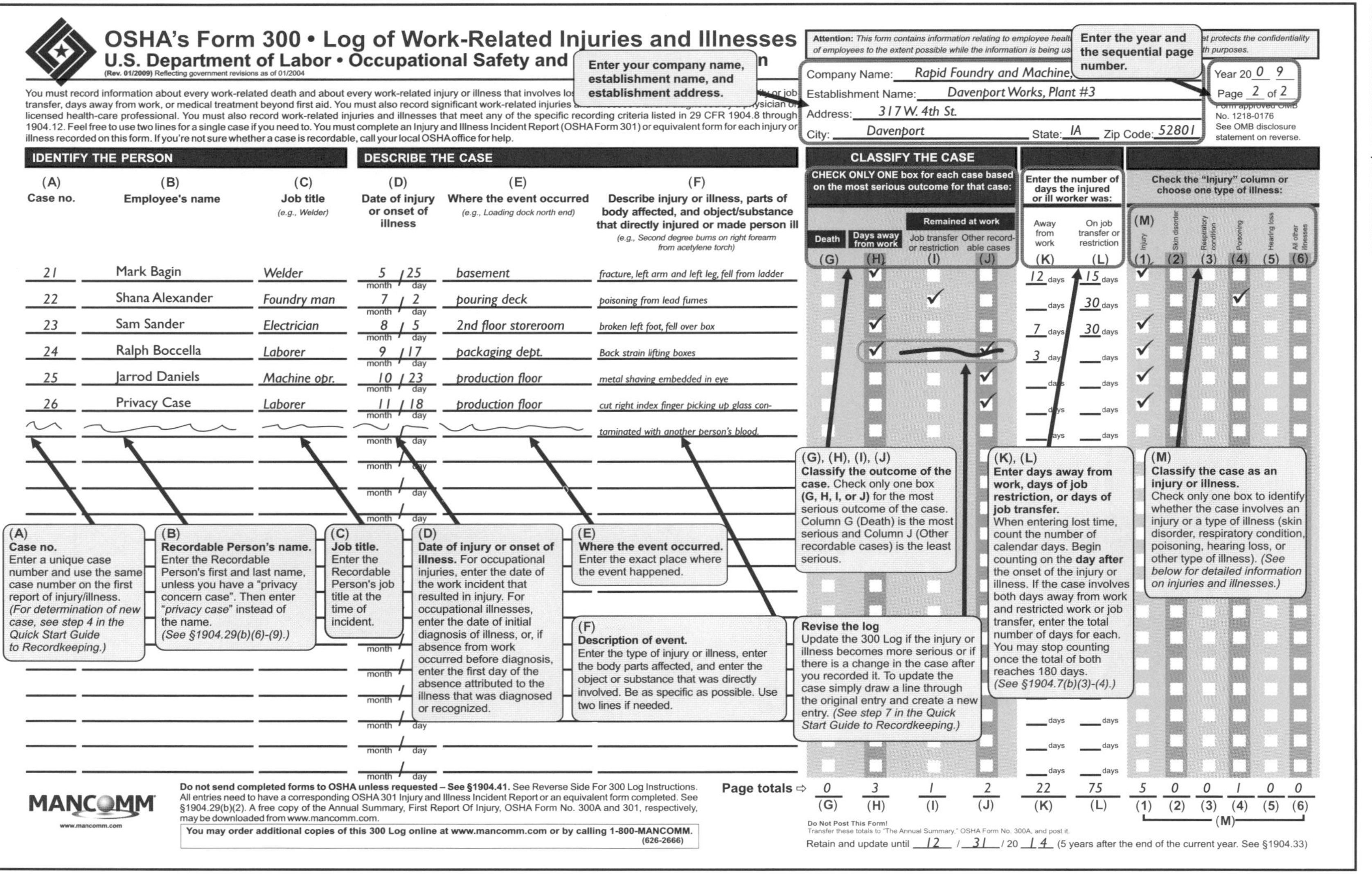

OSHA's Form 300 • Log of Work-Related Injuries and Illnesses
U.S. Department of Labor • Occupational Safety and Health Administration
(Rev. 01/2009) Reflecting government revisions as of 01/2004

You must record information about every work-related death and about every work-related injury or illness that involves loss of consciousness, restricted work activity or job transfer, days away from work, or medical treatment beyond first aid. You must also record significant work-related injuries and illnesses that are diagnosed by a physician or licensed health-care professional. You must also record work-related injuries and illnesses that meet any of the specific recording criteria listed in 29 CFR 1904.8 through 1904.12. Feel free to use two lines for a single case if you need to. You must complete an Injury and Illness Incident Report (OSHA Form 301) or equivalent form for each injury or illness recorded on this form. If you're not sure whether a case is recordable, call your local OSHA office for help.

Attention: This form contains information relating to employee health and must be used in a manner that protects the confidentiality of employees to the extent possible while the information is being used for occupational safety and health purposes.

Company Name: Rapid Foundry and Machine
Establishment Name: Davenport Works, Plant #3
Address: 317 W. 4th St.
City: Davenport State: IA Zip Code: 52801

Year 20 0 9
Page 2 of 2
Form approved OMB No. 1218-0176
See OMB disclosure statement on reverse.

(A) Case no.	(B) Employee's name	(C) Job title (e.g., Welder)	(D) Date of injury or onset of illness (month/day)	(E) Where the event occurred (e.g., Loading dock north end)	(F) Describe injury or illness, parts of body affected, and object/substance that directly injured or made person ill (e.g., Second degree burns on right forearm from acetylene torch)	(G) Death	(H) Days away from work	(I) Job transfer or restriction	(J) Other recordable cases	(K) Away from work (days)	(L) On job transfer or restriction (days)	(1) Injury	(2) Skin disorder	(3) Respiratory condition	(4) Poisoning	(5) Hearing loss	(6) All other illnesses
21	Mark Bagin	Welder	5/25	basement	fracture, left arm and left leg, fell from ladder		✓			12	15	✓					
22	Shana Alexander	Foundry man	7/2	pouring deck	poisoning from lead fumes			✓			30				✓		
23	Sam Sander	Electrician	8/5	2nd floor storeroom	broken left foot, fell over box		✓			7	30	✓					
24	Ralph Boccella	Laborer	9/17	packaging dept.	Back strain lifting boxes		✓		✓ (struck through)	3		✓					
25	Jarrod Daniels	Machine opr.	10/23	production floor	metal shaving embedded in eye				✓			✓					
26	Privacy Case	Laborer	11/18	production floor	cut right index finger picking up glass contaminated with another person's blood.				✓			✓					
Page totals						0	3	1	2	22	75	5	0	0	1	0	0

IDENTIFY THE PERSON — DESCRIBE THE CASE — CLASSIFY THE CASE
CHECK ONLY ONE box for each case based on the most serious outcome for that case.
Enter the number of days the injured or ill worker was:
Check the "Injury" column or choose one type of illness: (M)

Enter your company name, establishment name, and establishment address.

Enter the year and the sequential page number.

(A) **Case no.** Enter a unique case number and use the same case number on the first report of injury/illness. *(For determination of new case, see step 4 in the Quick Start Guide to Recordkeeping.)*

(B) **Recordable Person's name.** Enter the Recordable Person's first and last name, unless you have a "privacy concern case". Then enter "*privacy case*" instead of the name. *(See §1904.29(b)(6)-(9).)*

(C) **Job title.** Enter the Recordable Person's job title at the time of incident.

(D) **Date of injury or onset of illness.** For occupational injuries, enter the date of the work incident that resulted in injury. For occupational illnesses, enter the date of initial diagnosis of illness, or, if absence from work occurred before diagnosis, enter the first day of the absence attributed to the illness that was diagnosed or recognized.

(E) **Where the event occurred.** Enter the exact place where the event happened.

(F) **Description of event.** Enter the type of injury or illness, enter the body parts affected, and enter the object or substance that was directly involved. Be as specific as possible. Use two lines if needed.

(G), (H), (I), (J) **Classify the outcome of the case.** Check only one box **(G, H, I, or J)** for the most serious outcome of the case. Column G (Death) is the most serious and Column J (Other recordable cases) is the least serious.

Revise the log Update the 300 Log if the injury or illness becomes more serious or if there is a change in the case after you recorded it. To update the case simply draw a line through the original entry and create a new entry. *(See step 7 in the Quick Start Guide to Recordkeeping.)*

(K), (L) **Enter days away from work, days of job restriction, or days of job transfer.** When entering lost time, count the number of calendar days. Begin counting on the **day after** the onset of the injury or illness. If the case involves both days away from work and restricted work or job transfer, enter the total number of days for each. You may stop counting once the total of both reaches 180 days. *(See §1904.7(b)(3)-(4).)*

(M) **Classify the case as an injury or illness.** Check only one box to identify whether the case involves an injury or a type of illness (skin disorder, respiratory condition, poisoning, hearing loss, or other type of illness). *(See below for detailed information on injuries and illnesses.)*

MANCOMM www.mancomm.com

Do not send completed forms to OSHA unless requested – See §1904.41. See Reverse Side For 300 Log Instructions. All entries need to have a corresponding OSHA 301 Injury and Illness Incident Report or an equivalent form completed. See §1904.29(b)(2). A free copy of the Annual Summary, First Report Of Injury, OSHA Form No. 300A and 301, respectively, may be downloaded from www.mancomm.com.

You may order additional copies of this 300 Log online at www.mancomm.com or by calling 1-800-MANCOMM. (626-2666)

Do Not Post This Form! Transfer these totals to "The Annual Summary," OSHA Form No. 300A, and post it.

Retain and update until 12 / 31 / 20 14 (5 years after the end of the current year. See §1904.33)

(M)

Injury is any wound or damage to the body resulting from an incident in the workplace. *Examples:* Cut, puncture, laceration, abrasion, fracture, bruise, contusion, chipped tooth, amputation, insect bite, electrocution, or a thermal, chemical, electrical, or radiation burn. Sprain and strain injuries to muscles, joints, and connective tissues are injuries when they result from a slip, trip, fall, or other similar incidents.

Illnesses include:

Skin diseases or disorders that are caused by work exposure to chemicals, plants, or other substances. *Examples:* contact dermatitis, eczema, or rash caused by primary irritants and sensitizers or poisonous plants; oil acne; friction blisters; chrome, ulcers or inflammation of the skin;

Respiratory conditions are illnesses associated with breathing hazardous biological agents, chemicals, dust, gases, vapors, or fumes at work. *Examples:* silicosis, asbestosis, pneumonitis, pharyngitis, rhinitis or acute congestion; farmer's lung, beryllium disease, tuberculosis, occupational asthma, reactive airways dysfunction syndrome (RADS), chronic obstructive pulmonary disease (COPD), hypersensitivity pneumonitis, toxic inhalation injury, such as metal fume fever, chronic obstructive bronchitis, and other pneumoconioses;

Poisoning includes disorders caused by abnormal concentrations of toxic substances in blood, other tissues, other bodily fluids, or the breath from ingestion or absorption of toxic substances into the body. *Examples:* poisoning by lead, mercury, cadmium, arsenic, or other metals; poisoning by carbon monoxide, hydrogen sulfide, or other gases; poisoning by benzene, benzol, carbon tetrachloride, or other organic solvents; poisoning by insecticide sprays, such as parathion or lead arsenate; poisoning by other chemicals, such as formaldehyde;

Hearing loss is when an audiogram shows a change in hearing from the baseline audiogram of an average of 10 dB or more in either ear at 2000, 3000, and 4000 hertz and the Recordable Person's total hearing loss is 25 decibels (dB) or more above audiometric zero (also averaged at 2000, 3000, and 4000 hertz) in the same ear(s);

All other illnesses. *Examples:* heatstroke, sunstroke, heat exhaustion, heat stress, and other effects of environmental heat; freezing, frostbite, and other effects of exposure to low temperatures; decompression sickness; effects of ionizing radiation (isotopes, x-rays, radium); effects of non-ionizing radiation (welding flash, ultra-violet rays, lasers); anthrax; bloodborne pathogenic diseases, such as AIDS, HIV, hepatitis B or hepatitis C; brucellosis; malignant or benign tumors; histoplasmosis; or coccidioidomycosis.

MANCOMM
www.mancomm.com

6 88550 50652 6

Key Code: BXA

...d Illnesses
...nistration

...stricted work activity or job ...agnosed by a physician or ...d in 29 CFR 1904.8 through ...uivalent form for each injury or

Attention: *This form contains information relating to employee health and must be used in a manner that protects the confidentiality of employees to the extent possible while the information is being used for occupational safety and health purposes.*

Company Name: ______________________ Year 20___ ___

Establishment Name: ______________________ Page ___ of ___

Address: ______________________

City: ____________ State: ______ Zip Code: ______

Form approved OMB No. 1218-0176
See OMB disclosure statement on reverse.

(F) ...jury or illness, parts of ... and object/substance ...ured or made person ill (...ree burns on right forearm ...acetylene torch)	CLASSIFY THE CASE: CHECK ONLY ONE box for each case based on the most serious outcome for that case: Death (G)	Days away from work (H)	Remained at work: Job transfer or restriction (I)	Remained at work: Other record-able cases (J)	Enter the number of days the injured or ill worker was: Away from work (K)	On job transfer or restriction (L)	Check the "Injury" column or choose one type of illness: (M) Injury (1)	Skin disorder (2)	Respiratory condition (3)	Poisoning (4)	Hearing loss (5)	All other illnesses (6)
______	☐	☐	☐	☐	___ days	___ days	☐	☐	☐	☐	☐	☐
______	☐	☐	☐	☐	___ days	___ days	☐	☐	☐	☐	☐	☐
______	☐	☐	☐	☐	___ days	___ days	☐	☐	☐	☐	☐	☐
______	☐	☐	☐	☐	___ days	___ days	☐	☐	☐	☐	☐	☐
______	☐	☐	☐	☐	___ days	___ days	☐	☐	☐	☐	☐	☐
______	☐	☐	☐	☐	___ days	___ days	☐	☐	☐	☐	☐	☐
______	☐	☐	☐	☐	___ days	___ days	☐	☐	☐	☐	☐	☐
______	☐	☐	☐	☐	___ days	___ days	☐	☐	☐	☐	☐	☐
______	☐	☐	☐	☐	___ days	___ days	☐	☐	☐	☐	☐	☐
______	☐	☐	☐	☐	___ days	___ days	☐	☐	☐	☐	☐	☐
______	☐	☐	☐	☐	___ days	___ days	☐	☐	☐	☐	☐	☐
______	☐	☐	☐	☐	___ days	___ days	☐	☐	☐	☐	☐	☐
______	☐	☐	☐	☐	___ days	___ days	☐	☐	☐	☐	☐	☐
______	☐	☐	☐	☐	___ days	___ days	☐	☐	☐	☐	☐	☐
______	☐	☐	☐	☐	___ days	___ days	☐	☐	☐	☐	☐	☐
______	☐	☐	☐	☐	___ days	___ days	☐	☐	☐	☐	☐	☐
______	☐	☐	☐	☐	___ days	___ days	☐	☐	☐	☐	☐	☐
______	☐	☐	☐	☐	___ days	___ days	☐	☐	☐	☐	☐	☐
______	☐	☐	☐	☐	___ days	___ days	☐	☐	☐	☐	☐	☐
______	☐	☐	☐	☐	___ days	___ days	☐	☐	☐	☐	☐	☐
Page totals ⇨	___ (G)	___ (H)	___ (I)	___ (J)	___ (K)	___ (L)	___ (1)	___ (2)	___ (3)	___ (4)	___ (5)	___ (6)

(M) spans columns (1)–(6)

...ictions. ...ed. See ...ctively,

Do Not Post This Form!
Transfer these totals to "The Annual Summary," OSHA Form No. 300A, and post it.

Retain and update until ______ / ______ / 20 ___ ___ (5 years after the end of the current year. See §1904.33)

...OMM. ...666)

OSHA's Form 300 • Log of Work-Related Injuries a

U.S. Department of Labor • Occupational Safety and Health A

(Rev. 01/2009) Reflecting government revisions as of 01/2004

You must record information about every work-related death and about every work-related injury or illness that involves loss of consciousnes transfer, days away from work, or medical treatment beyond first aid. You must also record significant work-related injuries and illnesses tha licensed health-care professional. You must also record work-related injuries and illnesses that meet any of the specific recording crite 1904.12. Feel free to use two lines for a single case if you need to. You must complete an Injury and Illness Incident Report (OSHA Form 30 illness recorded on this form. If you're not sure whether a case is recordable, call your local OSHA office for help.

IDENTIFY THE PERSON			DESCRIBE THE CASE		
(A) Case no.	(B) Employee's name	(C) Job title *(e.g., Welder)*	(D) Date of injury or onset of illness	(E) Where the event occurred *(e.g., Loading dock north end)*	Describ body affec that directly *(e.g., Seco*
			month / day		
			month / day		
			month / day		
			month / day		
			month / day		
			month / day		
			month / day		
			month / day		
			month / day		
			month / day		
			month / day		
			month / day		
			month / day		
			month / day		
			month / day		
			month / day		
			month / day		
			month / day		
			month / day		
			month / day		

www.mancomm.com

Do not send completed forms to OSHA unless requested – See §1904.41. See Reverse Side For 300 Log Instru All entries need to have a corresponding OSHA 301 Injury and Illness Incident Report or an equivalent form complete §1904.29(b)(2). A free copy of the Annual Summary, First Report Of Injury, OSHA Form No. 300A and 301, respe may be downloaded from www.mancomm.com.

U.S. Department of Labor OSHA Form No.

(Rev. 01/2009) Reflecting government revisions as of 01/2004

Step by Step

Quick Start Guide

1. Are You Required to Maintain a 300 Log?

Companies must keep records of all recordable occupational injuries and illnesses for each single business establishment that is expected to be in operation for one year or longer, unless the company-wide count of Recordable People (see Recordable People in Definitions section below) was 10 or fewer at all times during the last calendar year or your company is listed as a low hazard industry. *(See §1904.46 for definition of establishment. See §1904.1-.3 for information on partial exceptions. See Table §1904 Appendix A to Subpart B for the list of low hazard industries. See §1904.30(a) for multiple business establishments. See §1904.31 for definition of Recordable People.)*

2. Is this Person Recordable?

You record on the 300 Log the recordable injuries and illnesses of all Recordable People, *(see Recordable People in Definitions Section below). (See §1904.31 for information on Recordable People.)*

3. Is this Case Work-Related?

An injury or illness is considered work-related if an event or exposure in the workplace caused or contributed to the incident or significantly aggravated a preexisting condition. Injuries and illnesses that happen in the workplace are presumed to be work-related, unless there is an exception. *(See §1904.5(b)(2) for exceptions.)* The work environment includes not only the facility, but other areas where the company has sent at least one Recordable Person to work or be present. *(See §1904.5 for determination of work-relatedness. See §1904.5 (b)(1) for work environment. See §1904.5 (b)(4) for the definition of "significantly aggravated".)*

4. Is this a New Case?

A new case is when a Recordable Person suffers a new injury or illness. It is also a new case when a Recordable Person has fully recovered from the previous injury or illness, and a new event or exposure in the workplace causes the signs or symptoms to reappear. *See Step 7, if it is not a new case. (See §1904.6 for determination of new cases.)*

5. Is this Case Recordable?

A case is recordable on the 300 Log when it involves any of the following work-related incidents: death *(§1904.7(b)(2))*; days away from work *(§1904.7(b)(3))*; restricted work or transfer to another job *(§1904.7(b)(4))*; medical treatment, *(see Medical Treatment Exceptions in Definitions section below) (§1904.7(b)(5))*; loss of consciousness *(§1904.7(b)(6))*; significant diagnosis by a healthcare professional *(§1904.7(b)(7))*; needlestick or sharps injuries *(§1904.8)*; medical removal due to a specific OSHA Standard *(§1904.9)*; occupational hearing loss *(§1904.10)*; or tuberculosis *(§1904.11.)*

If you answered "yes" to all of the steps 1 through 5, record the case.

6. How to Record this Case:

Fill out a 301 Incident Report and enter the case on the OSHA 300 Log within seven (7) calendar days. A 301 Incident Report or equivalent form must be filled out for each entry on the 300 Log. Enter the days of away from work and the days of restricted work or job transfer on the 300 Log, do not count the first day of the incident. Restricted work activity is when either the company or a healthcare professional places a work restriction on a Recordable Person or keeps a Recordable Person from performing the same duties and working the full workday that was scheduled before the incident. *(See §1904.29 for information on forms. See §1904.29(b)(4) for equivalent forms. See §1904.7(b)(4) for restricted work.)*

7. How to Update this Case: (when required)

If there are changes to the case during a five year period, you must update the original 300 Log entry to show these changes. Also enter on the original 300 Log any newly discovered recordable injuries or illnesses. *(See §1904.33(b) for information on updating records.)*

8. How to Complete, Post, and Retain Records:

At the end of the calendar year, total each column on the 300 Log, and then use this information to fill out a Form 300A, Summary of Work-Related Injuries and Illnesses. The 300A must be signed and certified by the highest-ranking company official at that facility or his/her immediate supervisor. Post the 300A Summary in that facility from February 1 to April 30 of the following year. Keep all 301 Reports, 300 Logs, and 300A Summaries for 5 years. *(See §1904.32 for information on annual completion of 300 Log and 300A Summary. See §1904.32(b)(5)-(6) for information on posting the 300A Summary. See §1904.33(a) for information on retention.)*

References and Definitions

Fault *(§1904.0)*
Cases listed on the 300 Log are not necessarily eligible for worker's compensation; nor does it mean that the company or Recordable Person was at fault or that an OSHA standard was violated.

Location *(§1904.35(b)(2)(iii); 1904.40)*
You may prepare and maintain this 300 Log at another facility; however, you must be able to show the OSHA Inspector these records within four business hours. When a worker or former worker requests these records, you must give them the records, with some limitations, by the end of the next business day.

First Report of Injury/Illness *(§1904.29(b)(3))*
You must complete a 301 Incident Report or equivalent form for each recordable case entered on the 300 Log. Many employers use an insurance form instead of the OSHA 301 Incident Report, or supplement an insurance form by adding any additional information required by OSHA.

Recordable People *(§1904.31)*
Recordable people are workers on your payroll and anyone else you supervise on a day-to-day basis who are not on your payroll.

Medical Treatment Exceptions
(are not considered recordable)
(§1904.7(b)(i), §1904.7(b)(ii))

1. A visit to a doctor or health-care professional only for observation or counseling; or
2. Diagnostic procedures such as x-rays, blood tests, and taking prescription medications used only for diagnostic purposes; or
3. The following procedures:
   - Taking non-prescription medication at non-prescription strength;
   - Receiving a tetanus shot;
   - Cleaning, flushing, or soaking wounds on the surface of the skin; using wound coverings, like bandages, Band-Aids™, gauze pads, etc., or using SteriStrips™ or butterfly bandages;
   - Using hot or cold therapy; using any non-rigid means of support, such as elastic bandages, wraps, non-rigid back belts, etc.; using temporary immobilization devices while transporting an accident victim (splints, slings, neck collars, or back boards);
   - Drilling of a fingernail or toenail to relieve pressure, or draining fluids from blisters;
   - Using eye patches; removing foreign bodies from the eye using only irrigation or a cotton swab;
   - Removing splinters or foreign material from areas other than the eye by irrigation, tweezers, cotton swabs, or other simple means;
   - Using finger guards; using massages; or drinking fluids to relieve heat stress.